Elements of
MICROPALAEONTOLOGY

Microfossils – Their Geological and Palaeobiological Applications

Gérard Bignot
(Académie de Paris, Laboratoire de Micropaléontologie)

A member of the Kluwer Academic Publishers Group
LONDON/DORDRECHT/BOSTON

First published in 1985 by
Graham & Trotman Limited
Sterling House
66 Wilton Road
London SW1V 1DE

French edition published by
Bordas Dunod, Paris
*Les Microfossiles: Les différents groupes
Exploitation paléobiologique et geologique*
© Bordas, Paris, 1982

British Library Cataloguing in Publication Data

Bignot, G.
 Elements of micropalaeontology : the microfossils –
 their geological and palaeobiological applications
 1. Micropalaeontology
 I. Title
 560   QE719

ISBN 0-86010-490-7

© Graham & Trotman, 1985
Reprinted 1994

This publication is protected by international copyright
law. All rights reserved. No part of this publication
may be reproduced, stored in a retrieval system, or transmitted
in any form or by any means, electronic, mechanical, photo-
copying, recording or otherwise, without the prior permission
of the publishers.

Phototypeset in Great Britain by Bookworm Typesetting, Manchester
Printed and bound in Great Britain by Watkiss Studios Limited,
Biggleswade, Beds.

PREFACE

Micropalaeontology, or the scientific study of microfossils, features prominently both in palaeontology and geology and its teaching has been developed through higher degree courses in Natural and Earth Sciences. Moreover, it forms an integral part of the training of professional geologists, as well as teachers in high schools and colleges.

Although recently our knowledge has increased dramatically, the information is not easily accessible; in fact it has resulted in numerous unco-ordinated studies. Admittedly, there have been several recent although incomplete studies published in English which are of good quality; but a simple and relatively exhaustive study of micropalaeontology has never been published. This text aims primarily to present the reader – palaeontologist, geologist or simply amateur – with a synopsis of the current state of knowledge.

Because of its nature and its limited context, such a project demands somewhat arbitrary choices and inevitably contains gaps and simplifications which will leave certain of my specialist colleagues dissatisfied. They would be misguided, moreover, in trying to find unpublished facts or original ideas. Above all, I have forced myself to be succinct and I hope the reader of this small book will discover the essential character of micropalaeontology, and a guideline as to the place of micropalaeontology in geology and palaeontology.

The many illustrations complete the book. These are deliberately precise, and serve to elucidate numerous points. It is thus I have summarized the systematics of the different groups in one or several figures which show at a glance, what could not be explained in words without lengthy and tedious detail. Some drawings are original; most of the others are copies or schematics of drawings published in various published works.

The continuous developments in their subject have led micropalaeontologists to consider the purpose of their study from various standpoints. This has resulted in the somewhat decompartmentalized division of the subject into several specialized fields. Even if I am not familiar with all of these I believe I have acquired enough direct knowledge about most to write about them. I must confess, however, that my personal contribution is of modest size when compared with the mass of documentation provided by conversations between colleagues, participation in symposiums and reading books and specialist articles. This disproportion must be emphasized especially since it was impossible to name all the authors asked to cooperate.

Finally this is a good opportunity for me to express my deepest gratitude to my teachers, the late Jean Cuvillier and Professor Louis Dangeard, as well as to all those who have helped and encouraged me in my work on several accounts. I would like to express my gratitude to Professor Charles Pomerol for encouraging me to undertake this work and to Professor Jean Dercourt whose help was vital for the correction and publication of the book. I also wish to convey my special thanks to Professor Madeleine Neumann, Misses Marie-Christine Janin and Edwige Masure, Messrs. Gérard Breton, Patrick deWever, Claude Guernet, George Lachkar and Lucien Lezaud. I also wish to add the names of Aget Language Services for the translation and Dr. Sally Radford, who acted as scientific reader for the English edition. I hope they all accept my gratitude and friendship.

Gérard Bignot
Laboratory of Micropalaeontology &
Laboratory of Comparative Stratigraphy
Department of Sedimentary Geology
University Pierre-et-Marie Curie, Paris

FOREWORD

If any beach sand is examined through a lens, it will soon be discovered that, amongst the grains of quartz and other minerals, there are very many small shells. These may include the juveniles of molluscs and other, potentially large, marine organisms; but there will also be innumerable minute shells of organisms which are never to grow larger than the grains of the enclosing sand. They will include the bivalved carapaces of microscopic arthropods (the ostracoda), the coiled, chambered shells of protozoans (the foraminifera), fragments of coralline algae, and many other skeletal kinds. In some tropical beaches, behind coral reefs, the sands are composed almost entirely of these microscopic, biogenic remains. Out to sea, far beyond the beaches, on the seabed of the continental shelf and slope, an even greater diversity of microscopic skeletons may accumulate in the sediment, accompanied by the virtually indestructible pollen and spores of land plants, blown there by the winds. On the floor of the abyssal ocean, there accumulates a grey, sticky ooze, composed almost entirely of the remains of planktonic unicellular plants and protozoans, tens of thousands of their skeletons being preserved in each gram of sediment. All of these are potential microfossils. After the passage of geological time, they form part of the sedimentary rocks, and, sometimes, almost wholly compose them.

Quantitatively, microfossils form by far the greatest part of the palaeontological record. Their frequent abundance in sedimentary rocks – from past marine, estuarine and even freshwater environments of deposition – make them easy to collect and economical to transport to a laboratory. Their microscopic size enables quantities of them to be retrieved from very small rock samples – even from the fragmentary rock cuttings flushed up from the drilling bit in an exploratory well. Their evolution has ensured that successive rock strata contain the remains of recognisably distinct species, enabling the geologist specialising in biostratigraphy to identify the ages of rock samples, and to correlate stratum with stratum, by the microfossil assemblages they contain.

The different groups of microscopic fossils may vary in relative abundance in sedimentary rocks which are of the same age, because the various organisms flourished in different depositional environments. The freshwater sediments of ancient lakes and deltas can yield diatoms, the pollen and spores of terrestrial and aquatic plants, and characteristic genera of ostracods. Seawards, the more brackish sediments of the delta front, lagoon or tidal estuary formed sediments containing characteristic admixtures of these together with salinity tolerant species of foraminifera and other microfossils which could live where the sea and land met. On the open shore, the coastal seabed and the continental shelf, fully marine assemblages of benthonic foraminifera, with ostracods and other small metazoa dominate, with the contribution by estuarine remains being much reduced. In the deep ocean, planktonic groups (calcareous nannoplankton, the globigerinacean foraminifera, diatoms, or radiolaria) become abundant, often swamping all other microfossils by their numbers. Thus, the palaeoenvironment under which the sediment was deposited can be determined, and palaeofacies, palaeogeographic and palaeobiographic maps may be compiled. Even palaeoclimatic and palaeoceanographic changes can be recognised and plotted, to aid the elucidation of the geological history of the world, from areas of the scale of single oilfields to that of the globe itself. By their abundance, diversity and environmental implications, the microfossils provide the palaeontologist with the most complete fossil record of the evolution of life which can be obtained.

The theoretical and practical importance of microfossil studies has grown with each decade since they were first employed in subsurface petroleum exploration, over 60 years ago. Today, the petroleum exploration companies, national geological surveys, geological exploration consultancies and the universities together employ more scientists engaged in micropalaeontology than in all other aspects of palaeontology combined. It has become increasingly necessary to introduce undergraduate students of geology to the potential applications of the study of microfossils, and for practising geologists to be aware of them. This book sets out to form such an introduction.

Nowadays, practising micropalaeontologists tend to specialise, in order to attempt to keep abreast of the advances being made in the study and application of particular groups of microfossils alone. Some may confine their attention to palynology, others to the coccoliths, and others to ostracods or foraminifera. With the last group, research has advanced so rapidly that it is not uncommon to deal with specialists in the planktonic foraminifera only, or in other equally restricted foraminiferal categories. It is rare to find an author who can write authoritatively, not only from his teaching and practice but from his own, original research, upon the broad spectrum of microfossils as a whole. Gérard Bignot is able to do this. He has researched, and has published his researches, on the systematics and ecology of 'large' and 'small' foraminifera, of ostracods, calcareous algae, calpionellids, pithonellids, diatoms and even of fossil bacteria. He has studied the role that these and fragmentary molluscs, corals, echinoderms and other metazoan skeletons play in the composition and 'microfacies' of sedimentary rocks, and how the extremely small 'nannofossils' (coccoliths and their allies) participate in the genesis of chalks, of diatomites and even of the ferromanganese nodules of the deep ocean. Bignot has also published extensively on the biostratigraphic, palaeoenvironmental, palaeogeographic and palaeobiological implications of his microfossil researches, especially with reference to western Europe and to the Mediterranean areas, and he is an active member of the International Subcommission on Palaeogene Stratigraphy, of the Association des Géologues du Bassin de Paris (of which he was president in 1978), and of the Service de la Carte géologique de France. He is also Editor of the international periodical, *Revue de Micropaléontologie*. From his laboratory at l'Université Pierre-et-Marie Curie, in Paris, his work continues, both in original research and in teaching. With this book, the breadth of his vision, and some of the fruits of his original investigations, should reach a yet wider public, and the significance of the study of microfossils should become even more thoroughly realised.

F.T. Banner
University College London

CONTENTS

Preface – Prof. Gérard Bignot ix

Foreword – Prof. F.T. Banner xi

PART ONE: MICROFOSSIL GROUPS

Chapter 1 Introduction

    1. Micropalaeontology and its Purpose: Microfossils   1
    2. Brief historical survey   2

Chapter 2 Collection, Preparation, Observation and identification of Microfossils

    1. Collection   5
    2. Preparation   7
    3. Observation   10
    4. Description and Identification   15

Chapter 3 Foraminifera

    1. Living foraminifera   19
    2. The test and its fossilization   23
    3. Systematic survey   30
    4. Foraminifera throughout geological time   37

Chapter 4 Ostracods

    1. Living ostracods   43
    2. The carapace and its fossilization   45
    3. Systematic survey   49
    4. Ostracods throughout geological time   51

Chapter 5 Calpionellids and Related Microfossils

    1. Calpionellids   56
    2. Calpionellomorphs   59
    3. Calcispheres   59

Chapter 6 Mineralized Plant and Animal Remains

1. Calcareous algae	61
2. Pteropods and tentaculitids	65
3. Isolated organic elements	68
4. Skeletal fragments	68

Chapter 7 Calcareous Nanofossils

1. Living coccolithophores	72
2. The systematic classification of coccoliths and its problems	74
3. Coccoliths throughout geological time	78
4. Some other calcareous nanofossils	80

Chapter 8 Siliceous Microfossils

1. Polycystine Radiolarians	83
2. Diatoms	89
3. Minor groups	94

Chapter 9 Conodonts

1. General organization	99
2. Systematic affinities and biological significance	102
3. Conodonts throughout geological time	104

Chapter 10 Palynology

1. Palynology and palynomorphs	106
2. Spores and pollen	107
3. Dinoflagellates	115
4. Acritarchs	121
5. Chitinozoans	123
6. Minor groups	125

Chapter 11 The position of Microfossils in the Systematic Classification of the Living World   129

PART TWO: GEOLOGICAL AND PALAEOBIOLOGICAL APPLICATIONS OF MICROPALAEONTOLOGY

Chapter 12 Microfossils in the Environment of Preservation

1. From living organism to microfossil	133
2. Deposition	138
3. Microfossils in deposits	142

Chapter 13 Microfossils – The Key to Biological Problems

1. From ecology to palaeoecology	145
2. Species and speciation	153
3. Evolutionary trends	159
4. Microfossils and the origins of life	163

Chapter 14 Microfossils as a source of sediments

1. Lithogenesis through bioclastic accumulation	169
2. Lithogenesis through the concentration of amorphous organic substances	172
3. The link betwen lithogenesis and microbiotic activity	174

Chapter 15 Microfossils – chronometers of the Phanerozoic

1. Microfacies	177
2. Biozones and biozonation	178
3. Biostratigraphy, chronostratigraphy, magnetic reversals and radiometric dating	182

Chapter 16 Microfossils as palaeoenvironmental and palaeogeographic indicators

1. From palaeoecology to the reconstruction of palaeoenvironments	187
2. Microfossils – evidence of sea-floor spreading	197
3. From palaeobiogeography to global palaeogeography	200

Chapter 17 General conclusion	206
Acknowledgements	209
Index	211

Part 1
Microfossil Groups

Chapter 1

Introduction

1. MICROPALAEONTOLOGY AND ITS PURPOSE: MICROFOSSILS

Micropalaeontology is concerned with **microfossils and nanofossils** (the latter being smaller than 50 μm)[1], the study of which must, of necessity, be carried out using the light or electron microscope. These fossils consist of:

- the remains of unicellular and multicellular microorganisms (**microbiota**); and
- the dissociated elements and skeletal fragments of macroorganisms (**macrobiota**).

The hard parts of living organisms that may be fossilized vary in their chemical nature: some are formed from silica, carbonates or phosphates of calcium; others are composed of non-mineralized organic elements.

It follows from the above definition of microfossils that all groups of plants and animals with a fossil record fall within the field of micropalaeontology, regardless of initial size. Moreover, the methods and goals of palaeontologists do not depend on the dimensions of the materials that they study. It might be asked whether micropalaeontology should exist as an independent discipline. For many years, its distinguishing characteristic was the obligatory resort to the microscope. This, however, is a distinction that has become increasingly blurred with the quest for ever finer morphological and microstructural detail by palaeobotanists, invertebrate palaeontologists and even vertebrate palaeontologists.

---

[1] From the Greek *nanos* or *nannos* (= very small). The term was proposed and defined by G. Deflandre (1959). Although the Académie des Sciences in Paris has recommended that it should be written 'nannofossil', it seems preferable to use the spelling 'nanofossil', keeping the same prefix as for units of physical measurement (nanometre, nanosecond, etc).

Micropalaeontology is not, therefore, a well-defined discipline, although for traditional reasons it is often considered as such. Foraminifera, ostracods, coccoliths, dinoflagellates and conodonts have always been regarded as microfossils but bryozoans, though apparently possessing all the right characteristics, are studied by specialists who do not consider themselves as micropalaeontologists. The place of the calcareous algae and the tentaculitids, to cite only two examples, is even more uncertain.

Tradition alone would not have been enough to ensure the independence of micropalaeontology if microfossils had not been of considerable practical and economic interest. Macrofossils are rare in outcrops and still more so in core samples and rock cuttings. By contrast, microfossils may be found in virtually every sedimentary rock sample. As they are generally well preserved and often abundant in even small volumes of sediment, they are used as marker fossils in oil prospecting. Every oil-exploration company thus uses the services of a micropalaeontology laboratory at some time.

On the whole, micropalaeontologists tend to form a group somewhat apart from the main body of palaeontologists. More often they work in applied geology (prospecting for oil and water, assessing the foundations for civil engineering works, investigating mines, etc.) than as specialists attached to university laboratories or involved in fundamental research. Moreover, those that do research have frequently been employed in industry or they maintain a close relationship with colleagues in the 'applied' branch.

As micropalaeontology is a vast subject, microfossils being numerous and varied, the goal of the micropalaeontologist is threefold:

- To study the fossils in terms of their morphology, microstructure, chemical and mineralogical components.
- To attempt to classify them, to discover their origins and systematic affinities.
- To strive, finally, to determine their role in petrogenesis, their stratigraphic importance, and their palaeobiological interest.

Nowadays it is extremely difficult for one researcher to master the whole field of micropalaeontology. Thus, micropalaeontologists must restrict themselves to a group, or part of a group that has been systematically delimited in terms of geography or stratigraphy.

2. BRIEF HISTORICAL SURVEY

Because of their large size, the nummulitid foraminifera (e.g. *Nummulites*) were the first microfossils to be noticed. Although he failed to discern their true significance, Strabo (58 BC – AD 25) noted their presence in the limestones used to build the Egyptian pyramids.

With the improvement of the magnifying glass, and eventually with the invention of the microscope, it became possible to study microorganisms and tiny 'petrificata' that are invisible to the naked eye. U. Aldrovandri (1522–1607) and R. Hooke (1635–1703) were the first to perceive foraminifera. The most celebrated micrographer of the age, however, was the Delft draper, A. van Leeuwenhoek (1632–1723), who revealed to his contemporaries the wonders of the microscopic world. At the time of the publication in 1758 of the tenth edition of *Systema Naturae*, the work of the Swede Linnaeus, no more than a score of species of foraminifera were known and these were attributed to the genera *Nautilus* and *Serpula*.

The first half of the nineteenth century was marked by the discovery of almost all the microfossil groups. It was often the case, however, that their systematic affinities and even their organic origin remained unknown. When A. d'Orbigny (1802–1857) began his career as a systematicist by proposing a classification for foraminifera, he thought that he was dealing with tiny cephalopods. Only later was it realized that they were quite separate microorganisms. The observation of microfossils in indurated rocks using thin sections began in 1849 with H.C. Sorby (1826–1908). The true founder of micropalaeontology, however, seems to have been C.G. Ehrenberg (1795–1876) who understood the rock-forming (lithogenic) role of the microfossils depicted in his *Mikrogeologie* completed in 1854.

As the number of species of microfossils continued to grow, it was not long before their stratigraphical importance began to be suspected. Already by 1823, *Nummulites* had been used stratigraphically by A. Brongniart. Not long afterwards the small(er) foraminifera were similarly employed. In 1850, E. Forbes proposed that the Purbeck Beds should be zoned on the basis of the ostracods they contained. At last, in 1874, there came the first utilization of foraminifera in the stratigraphical interpretation of a drilling sample: W. Dames and L.G. Bornemann proved that drilling near Greifswald (GDR) had penetrated Turonian rock. As the twentieth century dawned, the development of micropalaeontology accelerated:

- 1916: beginning of teaching as a specialist subject in several American universities.
- 1919: creation by the Humble and Rio Bravo Oil Co. of the first laboratory for 'applied' micropalaeontology with a staff of three researchers.
- 1923: foundation by J.A. Cushman (1881–1949) of the Laboratory for Foraminiferal Research at Sharon, Massachusetts, USA. For a quarter of a century, this was the headquarters of micropalaeontology.
- 1925: beginning of publication of the first periodical devoted entirely to microfossils.

During this period, micropalaeontologists were almost exclusively concerned with foraminifera. Other microfossils were studied only by isolated researchers such as the Frenchman L. Cayeux (1864–1944), who attempted to show their petrogenetic activity, and G. Deflandre (1897–1973) who specialized in fossil flagellates. The study of fossil spores and pollens (palynology) also emerged in this time. The study of Quaternary peats led the Swede, L. von Post (1884–1951), to develop pollen diagrams that he used in 1916 and later applied them, around 1930, to pre-Quaternary terrains.

Particularly after 1945, the growing need for hydrocarbons led to a rapid development of micropalaeontology. In France the impetus was provided by J. Cuvillier (1899–1969). It is now a science that is practised in almost every country in the world.

As the subject grew, the number of micropalaeontologists increased to around 12 000 and professional journals multiplied with thousands of articles being published every year. Although this flood of publications is deplored by some, nevertheless it remains a reflection of the vitality of the discipline.

BIBLIOGRAPHY

The following books deal with all, or almost all, microfossils:

M.F. Glaessner, *Principles of Micropalaeontology*, 2nd edn (Hafner, New York, 1963); D.J. Jones, *Introduction to Microfossils* (Harper Brothers, New York, 1956); V. Pokorny, *Principles of Zoological Micropalaeontology*, 2 vols (Pergamon Press, Oxford, 1963/5; 1st edn in Czech in 1954; 2nd edn in German in 1958). These books are now out of date and no longer present a panorama of modern micropalaeontology.

A.T.S. Ramsay (ed.), *Oceanic Micropalaeontology*, 2 vols (Academic Press, London and New York, 1977) is a work for specialists. B.U. Haq & A. Boersma (eds), *Introduction to Marine Micropalaeontology* (Elsevier, New York, 1978) and M.D. Brasier, *Microfossils* (Allen & Unwin, London, 1980) are more accessible but still do not provide a complete coverage of micropalaeontology.

Articles dealing with microfossils are published in several periodicals. Some of these are specialized and will be mentioned later. Others are more general, e.g. *Micropalaeontology*, New York (since 1955, follows on from the *Micropalaeontologist*, 1947–54); *Voprosi Micropalaeontologia*, Moscow (1956–); *Revue de Micropaléontologie*, Paris (1958–); *Cahiers de Micropaléontologie*, Paris (1965–); *Revista española de Micropaleontologia*, Madrid (1969–); *Utrecht Micropalaeontological Bulletin*, Utrecht (1969–); *Marine Micropalaeontology*, Amsterdam (1976–), and *Journal of Micropalaeontology*, London (1982–).

Chapter 2

Collection, Preparation, Observation and Identification of Microfossils

1. COLLECTION

Preliminary Remarks

In the field, micropalaeontologists begin working at the level of the outcrop with a detailed survey of the section complemented by annotated diagrams and photographs. Locations from which samples have been taken must be carefully noted. If a borehole has been drilled, the precise depth of the samples recovered must be given.

Microfossils are generally not visible to the naked eye but the larger ones (especially foraminifera) can be seen with the aid of a magnifying glass. They show up well on freshly split rock when this is moistened.

With a little experience, it is possible to predict the micropalaeontological content of a sediment. The presence of macrofossils is an interesting indicator although neither necessary nor sufficient. All rocks of sedimentary origin may contain microfossils but the numbers, variety and state of preservation vary greatly depending on the nature (Fig. 2.1), age and origin of the rocks. Graded bedding and sedimentation marks must be observed as they condition the distribution of microfossils in the sediment.

The taking of samples comes only after a series of observations carried out at the outcrop. The micropalaeontologist must also be a geologist alert to the problems of sedimentary series, to the manner in which the rocks were laid down and to the stages of their diagenetic evolution.

Qualities of a Good Sample

For the micropalaeontologist, as for every geologist, the one indispensable tool is the hammer. Where there is a soil covering, a mattock is often useful and a drill sometimes indispensable.

The qualities of a good sample are threefold:

PART 1: MICROFOSSIL GROUPS

rocks \ microfossils	diatoms	calpionellids	chitinozoans	radiolarians	coccoliths	ostracods	conodonts	foraminifera	dinocysts and acritrarchs	spores and pollens
evaporites										●
dolomites					●	●	●	●	●	●
sands and sandstones			●			●	●	●	●	●
coal, lignite, etc.									●	●
jaspeites, lydites, flint and chert	●			●	●			●	●	●
limestones		●	●	●	●	●	●	●	●	●
marls and clays	●	●	●	●	●	●	●	●	●	●
metamorphic rocks: schists, phyllites, marbles								○	○	○ ○

● abundant ● rare ○ sporadic

Fig. 2.1 Average microfossil content in some types of rock. No consideration is taken of age or sedimentary origin (marine, lacustrine or other)

(1) It is *clean* if, before sampling, the surface material is removed because this is always weathered and contaminated by microorganisms (diatoms) or recent pollens. Particular care must be taken when loose sediments are being removed for palynological examination, especially during springtime when the air is full of pollen. A good method is to clear the surface of the outcrop and then drive in perpendicularly a clean metal tube. Alternatively, a large piece of rock can be wrapped up unbroken and the core removed in the laboratory just before treatment.
(2) It is *representative* and *complete* if samples are taken separately from every bank and bed in the section. The volume required is determined by the lithology: the poorer in remains the rock is presumed to be, the greater the volume. With a loose rock, 200 to 500 g is necessary for a complete study. Specialists in nanofossils can make do with a few grams. A small cube with sides of 4 to 5 cm is sufficient for hard rocks that are to be examined in thin sections.
(3) It is *identified* if it is hermetically sealed in plastic with all the references necessary for its identification written in indelible ink both outside and inside the container.

It cannot be stressed too often that good sampling is of fundamental importance; it is a task that micropalaeontologists must carry out for themselves. During fieldwork, they gather all the data necessary to pursue their project and to proceed with the palaeoecological and palaeogeographical interpretation. Knowing in advance the precautions demanded in sampling and proceeding in accordance with them, they can reduce to a minimum the risk of microfossils becoming mixed.

COLLECTION, PREPARATION, OBSERVATION AND IDENTIFICATION   7

2. PREPARATION

A detailed examination of laboratory procedures is beyond the scope of this book. Discussion of them will be restricted to indication of the more usual techniques and readers who wish to have further information are referred to the specialized works cited in the bibliography.

Whatever the techniques used, care and method are indispensable throughout the preparation of material for study. Contamination of samples, accidental mixes and errors in labelling always have unhappy consequences.

Mechanical Extraction Processes

These can be applied only to materials that are loose or no more than slightly indurated as the principle is to disaggregate the sediment by separating the grains according to size.

Rubbing

This is the quickest procedure for the preparation of nanofossils. A small fragment (1 to 2 mm^3) of rock that has been disaggregated in water is spread out on a slide. Good preparations are obtained by using the Lezaud method (Fig. 2.2.). Material can be observed dry, after drying in air, or in a liquid medium between slide and coverslip.

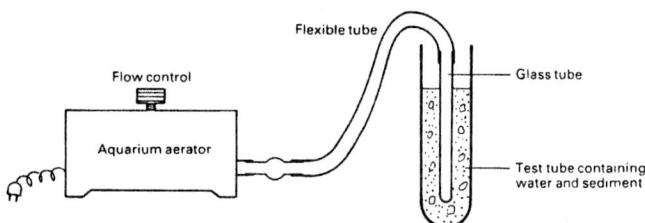

Fig. 2.2 Apparatus for the preparation of nanofossils; 2 to 3 hours of stirring in the current is generally sufficient (procedure was devised by Lezaud 1964, fig. 1)

Washing

This is the method most commonly used to extract microfossils that are larger than 100 μm. After drying, the rock is immersed in either pure water or in water to which a moistening agent (Teepol) has been added. Where necessary, it may be left to steep in a solution of 10 to 15% oxygenated water (H_2O_2) to 110 vol., neutralized with a few drops of ammonia. The disaggregated sediment is passed through a system of circular sieves (Fig. 2.3) with a metal base. From the top downwards, the arrangement is:

- A first (coarse) sieve with a mesh of 0.500 mm.
- A second (medium) sieve with a mesh of 0.160 mm.
- A third (fine) sieve with a mesh of 0.100 mm.

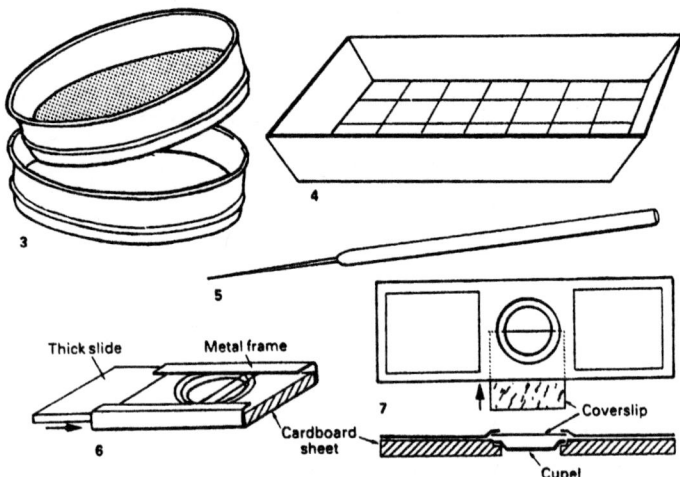

Figs 2.3–7 Material necessary for the washing and sorting of microfossils in unconsolidated rock
3 Sieve: usual diameter, 200 mm.
4 Dish with black bottom marked out in squares: usual dimensions 110 × 70 × 10 mm.
5 Mounted needle. 6–7 'Slides' (A. Francke, 1935):
6, German model (40 × 25 mm); 7, international model (75 × 25 mm)

A stream of water is run over the sediment, which is stirred gently with the fingertips. The operation is complete when the liquid leaving the column is clear. The residues are gathered by passing them into a cupel from the edge of the sieve with the help of a light trickle of water from underneath.

Each time they are used, the sieves are thoroughly scrubbed and then immersed for a few minutes in a solution with 5% methylene blue. The microfossils that remain caught in the mesh are thus stained blue and can be recovered in later washes.

With the microfossils captured by the sieves are other residues such as quartz, limestone fragments, grains of glauconite, etc. The materials must, therefore, be sorted. The dry residue is spread out on the bottom of a small dish with a black bottom (Fig. 2.4) and is examined under a binocular lens at magnifications of around × 25. The microfossils are removed using a fine brush or the point of a mounted needle (Fig. 2.5), the tip of which is inserted from time to time into Plastiline (a commercial modelling clay). They are then placed in the 'slides' shown in Figs 2.6 and 7.

Chemical Extraction Methods

Siliceous microfossils can be extracted from limestone gangue by the action of acids. However, conodonts, being composed of phosphates of calcium, are resistant to the action of weak acids. For this reason, they are extracted by immersing the crushed rock in a beaker filled with acetic acid (CH_3. COOH), monochloracetic acid (CH_2 Cl.COOH), or formic acid (H.COOH) diluted by 10 to 20%. The phosphate microfossils remain in the residue.

To extract non-mineralized microfossils, palynologists have devised techniques that require a well-equipped laboratory with ventilation hood, centrifuge, glassware, Teflon beakers, etc. The standard method for common rocks (sandstones, limestones, marls, schists, etc.) uses a 10-g sample and consists of the following stages:

- Crushing into fragments of less than 5 mm if it is a coherent rock.
- Destruction of any carbonate phase by the action of 50% hydrochloric acid (HCl).
- Attack on silica and silicates using 70% hydrofluoric acid (HF) (very dangerous!) over a period of 12 hours.
- Dissolution of the fluorosilicates formed using 50% HCl for 10 to 30 min.
- Selective oxidation of fine humic and carbonaceous particles using 10% KOH for 30 min.
- Elimination of pyrite and thinning out of palynological material through the brief (a few minutes) use of concentrated nitric acid (HNO_3).

Each phase in the process is followed by one or two rinses in water and then the use of the centrifuge. The residue is subsequently concentrated by sieving through fine-meshed (5 to 20 μm) nylon cloths that allow the passage of any remaining small particles. The sieves are used only once and are then disposed of. The deposit obtained is preserved in small tubes either in an aqueous medium or glycerized. The preparations are observed between slide and coverslip.

Thin Sectioning

The preparation of polished surfaces for hard rocks has largely been abandoned in favour of making thin sections. At least two sections are cut from each sample, one parallel and the other perpendicular to the stratification. The stages involved are as follows:

- Sawing of a section of rock bounded by two parallel plane faces.
- Polishing of one side using an abrasive (emery) moistened with water.
- Canada balsam or a synthetic resin (Araldite) is used to glue this side to a sheet of glass (usually 43 × 30 mm), which acts a mount.
- The other side is worn down until the rock becomes transparent (at a thickness of 30 to 50 μm).
- Balsam or resin is then used to cover this sliver of rock with a thin (0.1 mm) glass covering.

If the rock is ground too thin, the details of the microfossils will be obscured. For this reason, the thin sections used in micropalaeontology are thicker than ordinary petrographical sections: 30 to 50 μm instead of 25 μm. One effect of this is to heighten the polarization colours of the minerals: the quartzes are yellow, orange or even red.

The tools required can be quite simple: a metal saw, grindstone, wet and dry paper or an emery-faced glass block. The grade of emery must be adjusted to the different thicknesses of the section as it is being ground down: 200 μm at the beginning, 40 μm for finishing. Laboratories are equipped with more sophisticated apparatus: diamond-tipped circular saws, lapidaries' lathes with bronze face plates, 'truing' wheels and even specially constructed machinery.

3. OBSERVATION

Micropalaeontologists make their first observations in the field with the naked eye and the magnifying glass. On returning to the laboratory, they use the microscope.

Magnifying Glasses and Microscopes

At maximum accommodation (i.e. the minimum distance for clear vision, $\Delta = 25$ cm for the adult), the normal human eye is capable of distinguishing two points that are 0.075 mm apart (= 75 μm). This value is called the limit of resolution (d).

A **magnifying glass** gives a virtual image that is upright and enlarged. The magnification M of normal models is around 10. Given this value, the d_1 for a normal eye using a magnifying glass is:

$$d_1 = \frac{d}{M} = \frac{75}{10} = 7.5 \, \mu m$$

The optical (or **light**) microscope consists of two sets of lenses. The first is the objective lens, which gives a real image that is inverted and magnified. The second, or eyepiece lens, magnifies this image further. The d_m for this type of compound microscope is given by the equation:

$$d_m = 0.61 \, \lambda \, /(n \sin u)$$

Where λ is the wavelength of light (between 0.5 and 0.6 μm for white light); n is the index for a transparent medium separating the object from the objective ($n = 1$ for air in the case of a 'dry' lens); and u is half the aperture angle of a beam of light entering the objective.

It is possible to increase the value of $n \sin u$ (the numerical aperture) by using:

- An immersion objective in which a liquid of high refractive index is placed between the objective and the preparation. This may be water ($n = 1.33$) or, better still, cedarwood oil ($n = 1.52$).
- A condenser to increase the angle u.

In the best possible case, $d_m = 0.25$ μm, corresponding to a magnification of:

$$M = \frac{d}{d_m} = \frac{75}{0.25} = 300$$

In practice, magnifications are used that are four or five times greater than this value because observations at the limit of resolution are difficult. The enlarged images, however, do not show greater detail and tend to be blurred.

The optical microscope is used with direct light and the light rays received are transmitted rays (i.e. they have passed through the preparation). For this reason the preparation must be very thin: thin sections or microfossils smaller than 50 or 100 µm mounted between slide and coverslip.

The image that is produced is inverted, a phenomenon that makes for difficulty if the microfossil is to be moved or dissected. The image is therefore returned to the upright position by putting the light rays through four sets of reflections, using mirrors oriented in the appropriate planes. If two correcting microscopes are coupled and set at an angle of about 8° to the central axis of the instrument, the preparation can be viewed in relief. This stereomicroscope (H.S. Greenough 1897) is commonly called a **binocular lens**.

Because of the need for access to manipulate the object, there has to be a considerable working distance, a resolution limit of around 2 µm, and a relatively low magnification (from ×40 to ×120). In terms of field, this gives both good depth and area, and facilitates the preparation and examination of specimens. As both eyes are at work, fatigue is reduced and, at the same time, the equipment allows observation using both diascopy and episcopy (Fig. 2.8).

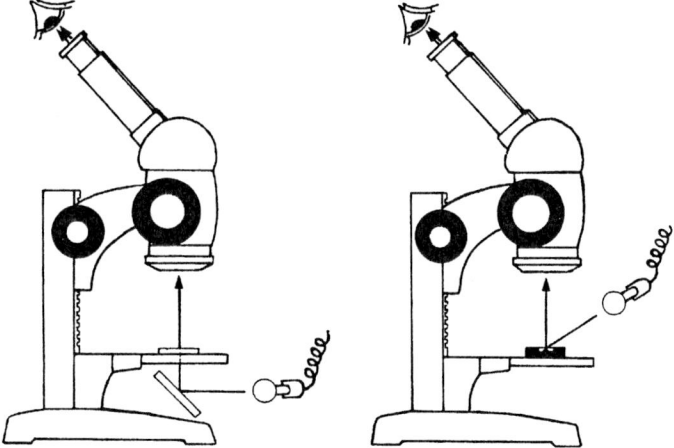

Fig. 2.8 The two ways of using the binocular lens. Left, transmitted light (diascopy): the microscope collects transmitted rays that have passed through the preparation. Right, reflected light (episcopy): the microscope collects rays reflected from the surface of the microfossil

The binocular lens is the tool most used by micropalaeontologists for routine work – sorting and observation of microfossils whether in thin sections or detached. The ordinary microscope tends to be used only when detailed examination is needed, particularly for nanofossils.

Up to about 1965, micropalaeontologists wishing to observe material at very high magnifications would resort to the **transmission electron microscope** (TEM). Although interesting results were obtained, this equipment is used nowadays only by palynologists. The reason for this is that the preparation of material is a very delicate process, requiring the making of ultrathin sections (<100Å) or surface replicas.

Once the **scanning electron microscope** (SEM) was perfected and brought on to the market, considerable progress became possible. The equipment was immediately seen to be indispensable so that, despite the high cost, many micropalaeontology laboratories – in industry or university – have at least one. The underlying principle (Fig. 2.9) of the SEM is that an object subjected to bombardment by electrons in a vacuum reacts electrically. Under the impact of the incident beam (or electronic probe), the sample gives off different types of

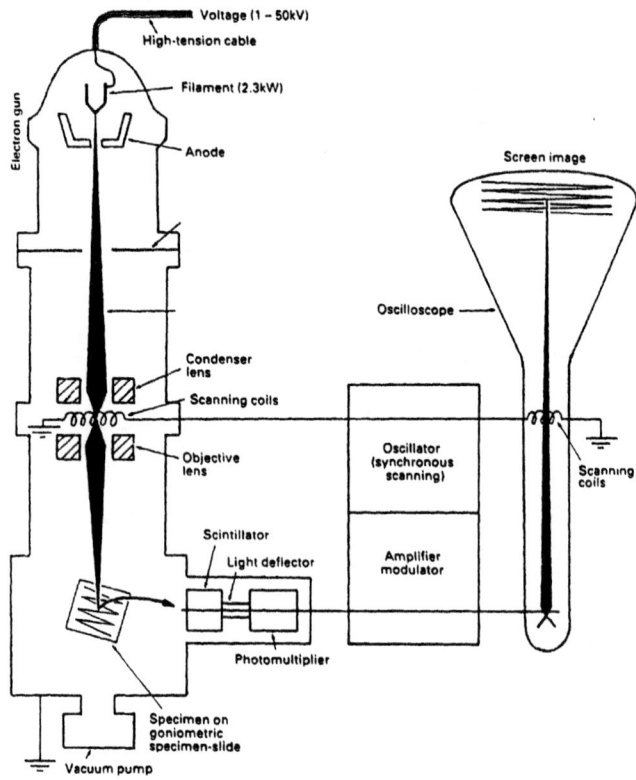

Fig. 2.9 Diagram of a scanning electron microscope.

COLLECTION, PREPARATION, OBSERVATION AND IDENTIFICATION 13

radiation, including secondary electrons and X-rays. The incident electrons are produced by a tungsten filament heated to incandescence. The beam thus produced is focused on the specimen by electromagnetic lenses. The diameter of the probe varies according to the model used: in the past, the range has been from 250 to 100 Å but recent equipment is capable of 50 Å.

In order to obtain an electronic image of the surface of the object, this is scanned with the probe. The electrons emitted by the surface are collected by a detector (scintillator) thus modulating the beam of a synchronous-scanning oscilloscope tube. At any one time, there is a correspondence between a point on the object and a point of the image on the screen. The intensity of emission of secondary electrons at any point on the object depends primarily on the incident angle of the probe. As the angle decreases, emission increases. The higher the power of emission of a point on the object, the more brilliant the corresponding point on the screen. Inclined surfaces, ridges and points are highlighted and the contrasts produce a three-dimensional effect. The image seen on the screen of the oscilloscope is a faithful reproduction of the surface of the object.

The resolution limit d_e of the SEM is equal to the area of emission of secondary electrons, an area almost identical to the surface of the object covered by the probe. The minimum magnification sufficient to show every detail is, using current equipment:

$$M = \frac{d}{d_E} = \frac{750\,000}{250 \text{ to } 100} = 3000 \text{ to } 7500.$$

As with optical microscopes, the practice is to use images with a magnification that is four or five times higher (i.e. from 10 000 to 30 000).

The material to be observed must be capable of undergoing dehydration and introduction into a high vacuum. However, preparation is extremely simple: the microfossil or scale of rock is stuck to the object-slide and its surface is rendered conductive by coating it through evaporation in a vacuum with a very thin film of gold or carbon.

The advantages of the SEM are as follows:

- The resolution limit is very low: 250 to 100 Å (ten times lower than for the best light microscope).
- Great depth of field providing a good stereoscopic view of the object.
- The material for study is easily prepared (dimensions can be as large as 1 cm) and the method is non-destructive.
- Possibility of varying the angle of observation (gyroscopic specimen stage).

In conclusion, therefore, it can be seen that micropalaeontologists have at their disposal a whole range of instruments of different capacities with which to observe microfossils.

Methods for Examining Microfossils

The complete morphological examination of a microfossil requires some five or six successive stages, two in the field and three or four in the laboratory (Fig. 2.10). The final stages after the preparation of the material obviously depend

	Equipment	Limit of resolution	Effective and normal magnification	Methods used to prepare material
In the field	Naked eye	~ 75 μm	1	No preparation
	Magnifying glass	~ 7.5 μm	~ 10–25	
In the laboratory — Routine	Binocular lens	~ 2 μm	~ 40–120	1. In reflected light: microfossils extracted by washing or chemical methods 2. In transmitted light: thin sections (microfacies)
	Light microscope	> 0.25 μm	~ 300 15–3250	1. Nanofossils and palynological microfossils in a liquid medium between slide and coverslip 2. Thin sections
In the laboratory — Detailed examination	SEM	250–50Å	3000–7500 15–50,000	1. Microfossils extracted by washing or chemical methods 2. Dry rubbing 3. Rock slides (nanofacies)
	TEM	5Å	> 150,000 200–500,000	Surface replicas and ultra-thin sections of microfossils

Fig 2.10 Stages in the morphological examination of microfossils. (Note: 1 mm = 1000 μm and 1 μm = 10 000 Å)

on the techniques used and on the size of the microfossils involved.

Once isolated from the sediment by washing or chemical extraction, large microfossils (>100 μm) are dried and examined, either as they are or after dissection, first under a binocular lens with direct lighting and then with the SEM. Those of smaller size, and nanofossils obtained by mechanical disaggregation or chemical extraction, are preserved in a liquid medium. They are examined between slide and coverslip under the light microscope or using the SEM after the evaporation of a suspension droplet on the object-slide.

The purpose of washing and chemical extraction is to isolate a fraction of the sediment characterized by either grain size or chemical composition. In either case, the micropalaeontologist has only a partial view of the mineralogical and palaeontological constituents of the sediment. Although all the constituents of the rock are there, they are dissociated and the micropalaeontologist cannot see their mutual relationships.

If, however, thin sections are studied under the binocular lens or light microscope, it is possible to view the sediment in its entirety and the constituents preserve their respective dispositions. This method provides a synthesis, giving an overall image of the rock that J. Cuvillier proposed should be called the **microfacies**. In reality it is no more than a facies examined at the magnifications permitted by the light microscope rather than by the naked eye. When a fragment of rock is examined in the SEM, a far more detailed overall image is produced; this is called the **nanofacies**.

Detection of Chemical and Mineralogical Compositions

Faced in the early days with the problems posed by the chemical and mineralogical compositions of their samples, micropalaeontologists could, at first, look for assistance only to the polarizing microscope and to a few chemical reactions. These simple and inexpensive procedures still render real service.

Already before the Second World War, however, X-ray diffraction methods, and particularly the powder diffraction method, were being applied to the mineralogical analysis of fossil shells. This made it possible to clear up some uncertainties: F.K. Mayer, for example, established in 1934 that the tests of foraminifera are, for the most part, calcitic.

In more recent years, micropalaeontologists have put to good use the arsenal of instruments developed by their colleagues in mineralogy and solid physics. Foremost among these is the **electronic probe microanalyser** (usually abbreviated to 'electron microprobe'). Analysis of the X-ray spectrum emitted by the sample under the impact of incident electrons provides information about its basic chemical composition. Depending on the method of detection used for the X-rays, the following is obtained:

- Either an image formed from points indicating the overall distribution of a selected chemical element (wavelength-dispersive spectrometer).
- Or a spectrum with points indicating the relative concentrations of different elements (energy-dispersive spectrometer).

The coupling of detectors such as these to an SEM makes possible the illuminating superimposition of data on the chemical elements over the secondary electron image. The preparation of material is identical to that required for the SEM, the only difference being that the surface for analysis must be almost plane and rendered conductive by coating with a thin carbon film.

Another technique involved in the study of microfossils is that which gives the carbonate isotope ratios $^{18}O/^{16}O$ and $^{13}C/^{12}C$. Techniques such as those just mentioned are not used only by micropalaeontologists. However, they are worthy of mention here not only because they are being used more frequently by micropalaeontologists, but also because they produce valuable results that raise great hopes for the future.

4. DESCRIPTION AND IDENTIFICATION

This section is concerned solely with the technical aspects involved in the identification of microfossils. The associated biological problems will be discussed later.

Description and Illustration: Synthesis of Observations

All the information gathered about a microfossil or microfacies is expressed in a description and one or more illustrations. No matter how precise the description, a drawing or a photograph always gives a clearer idea of the subject.

Drawings may be done free-hand or with the help of a *camera lucida*. Without going to the length of caricature, a good drawing should bring out the principal characteristics of the microfossil under consideration. Nowadays, photographs are preferred because they are quick and objective, and, more particularly, because the SEM takes excellent shots.

An illustration must always indicate the degree of magnification.

Identification

Description and illustration are a prelude to the next stage – the identification of the microfossil. At this point it receives its **binomen**.

It should be recalled here that, since Linnaeus, every species has been designated by two Latin names, which are written in italics in texts. The first name indicates the **genus** to which the **species** belongs and begins with a capital letter. The second, written all in lowercase, is the specific name. The name of the author(s) who named the species (its appellation) is generally included. A comma separates this from the year of the initial description:

e.g. : *Globigerina bulloides* d'Orbigny, 1826

If the authority is placed in parentheses, this indicates that the generic attribution of the species is not that which was originally proposed:

e.g. : *Globotruncana elevata* (Brotzen, 1934) was originally considered as belonging to the genus *Rotalia*

Where necessary, the name of a subgenus is placed in parentheses between the name of the genus and that of the species. A subspecies also has a trinominal nomenclature, its name being followed by those of the authors creating it:

e.g. : *Alveolina (Glomalveolina) primaeva* Reichel, 1936
Globorotalia cerroazulensis cunialensis Toumarkine & Bolli, 1970

Ideally, identification should involve a direct comparison of the samples being studied with those that served in the naming of the species. Unfortunately, however, the originals, where they have not disappeared, are rarely accessible but are scattered through various collections across the world. This being the case, it is necessary to refer to the publications in which the species were named, described and drawn for the first time. In principle, these documents are sufficient to characterize the species recorded and to establish whether or not an individual to be determined belongs to them. It is also possible to compare the individual with reference collections and to illustrations in works later than the original description. It must always be remembered, however, that the information contained in these documents may be erroneous or out of date. Another factor to bear in mind is that many species have several names, having been considered as new by authors who, in good or bad faith, neglected the publications of their predecessors. The real name is the one that was given first, but, in the interests of good mutual understanding and to broaden later conclusions, a list of all the different names attributed to a single species should be prepared: this is called a **synonymy**.

This bibliographical work is not peculiar to micropalaeontologists; the identification of a microfossil requires the same procedures as for a macrofossil.

It is no more difficult to determine a foraminiferid or a coccolith than an ammonite or a fish. Indeed, for two reasons, it may be easier:

- Microfossils are almost always more numerous and better preserved than macrofossils. This makes it easier to observe their morphology and to assess variations within a species.
- Bibliographical research is enormously simplified because the indexes and catalogues for all species of microfossils are relatively small in volume and are easy to consult. These important works are published by oil companies and public institutions. They contain all the original literature concerning a given group and are periodically updated. They are the envy of colleagues in macropalaeontology who do not possess documentation of this kind.

The work of identification requires documentation, rigour and prudence. If specimens are in a defective state and/or bibliographical data are inadequate, precise identification is impossible. In such a case, it is necessary to make do with an incomplete determination. This 'open' form is expressed by placing the abbreviation cf. between the generic and specific names, or by replacing the specific name by sp. On occasions, it may be necessary to stop at the level of the family, the order, or even the class or branch. It is far preferable to keep error at bay by giving an incomplete but prudent identification rather than one that is precise but erroneous.

Computers – the Solution for the Future?

For the time being, it is not possible to computerize the descriptions and illustrations necessary for palaeontological determinations. The main obstacle lies in the absence of homogeneity in the data. A coherent and universal system needs drawing up to codify the distinctive characteristics. All the known species will then have to be reviewed on this basis with statistical evaluation of intraspecific variation for each of them. This work is projected but is so vast that there is hesitation about undertaking it. Some interesting results, however, have already been obtained from partial attempts involving certain acritarchs and ostracods.

The creation of data banks governed by computers is a long-term undertaking but one that is full of promise. Indeed, only when such a system is put into effect by future generations will there be a solution to the problems brought about by the ever-growing numbers and diversity of publications.

CONCLUSION

The techniques involved in preparing microfossils and the equipment used for observation have been seen to be varied as they must meet the needs of:

- the size and chemical nature of the microfossils studied, and
- the properties and mineralogical nature of the fossil environment.

Readers will now have some inkling of the diversity of microfossils. Other aspects of this diversity will be brought out in the following chapters.

BIBLIOGRAPHY

A precise and complete description of the preparation of different microfossils is to be found in: *Techniques de Laboratoire en Géologie petrolière* (Technip, Paris, 1964) and *Méthodes modernes de Geologies de terrain: 3; Techniques d'echantillonage.* (Technip, Paris), 104pp. More theoretical, but rich in references, is B. Kummel & D. Raup (eds), *Handbook of Paleontological Techniques* (W.H. Freeman, San Francisco, 1965). For the collection, preparation and examination of calcareous microfossils (*not* nanofossils) see J.R. Haynes, *Foraminifera* Chap. 2 (Macmillan, London, 1981).

Many works deal with the description and use of the light microscope. The most useful are D.J. Jones, *Introduction to Microfossils* pp. 337-341, "Use and care of microscope" (Harper, New York, 1956) and G. Deflandre, *Microscopie pratique*, 2nd edn (Lechevallier, Paris, 1947).

Works concerning the SEM are still rare. A survey is given in J.-P. Eberhart, *Methodes physiques d'études des Minéraux et des Matériaux solides* (Doin, Paris, 1976). Micropalaeontological techniques peculiar to the SEM are described and commented on in the following:– W.W. Hay and P.A. Sandberg, 1967. The scanning-electron microscope: a major breakthrough for micropalaeontology. *Micropalaeontology* 13 (4), 407-418.

S. Kimoto and J.C. Russ, 1969. The characteristics and applications of the scanning electron microscope. *American Scientist* 57 (1), 112-133.

D.A. Walker, 1978. Preparation of geological samples for scanning electron microscopy pp. 185-192 *In: Scanning Electron Microscopy*, vol. 1; (SEM Inc., Illinois).

The rules for nomenclature are set out and explained in the *International Code for Zoological Nomenclature*, adopted at the Fifteenth International Congress in London, 1958 (last edn 1964) and in the *International Code for Botanical Nomenclature*, adopted at the Ninth International Congress in Montreal, 1959 (last edn 1969).

APPENDIX

Only for those who have the ambition to become palynologists is there any difficulty in setting up a small micropalaeontological laboratory. The basic material can be obtained from shops specializing in natural history. Although this last chapter may have given the impression that the use of the SEM is now the norm, it should not be thought that the light microscope is obsolete. If observations are carried out with rigour and perseverance, work with a modest binocular lens can be of real value.

If the student specializes in the microfossils of a given region and/or some restricted stratigraphic interval, bibliography should pose no insuperable problems. Information can be obtained from scientific libraries (natural history museums, universities, etc.) or, better still, from direct contact with professional micropalaeontologists who are, on the whole, understanding and responsive to the needs of beginners.

Chapter 3
Foraminifera

The order Foraminiferida – or foraminifera as they are informally called – forms the most important group of microfossils for two main reasons: first, they are abundant in rocks and there are numerous species which have given rise to a multiplicity of works over the years; second, they provide a valuable information in the dating of strata and the reconstruction of sedimentary environments.

1. LIVING FORAMINIFERA

General Organization

Foraminifera are unicellular organisms (Fig. 3.1) belonging to the rhizopod Protozoa (Protista). Their protoplasm, differentiated into **endoplasm** and **ectoplasm**, is emitted in the form of retractile **pseudopodia**, which are granular, anastomosing filaments. These are used in catching prey, in locomotion and in the creation of the skeleton.

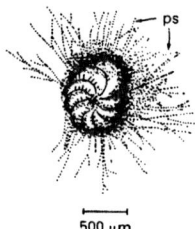

Fig. 3.1 Living foraminiferid: *Elphidium crispum* (Linnaeus, 1758); ps = pseudopodia (×15)

The feature that distinguishes foraminifera is the possession of a mineralized, intra-ectoplasmic skeleton or test formed from chambers that are interconnected by openings or **foramina**. The interior of the test is lined with a basal organic 'chitinoid' layer. This description should not prejudge its chemical nature which, though close to chitin, remains unknown.

Reproductive Cycle and Biology

It was not long before micropalaeontologists realized that a single foraminiferid species is often formed from two sorts of individuals with slightly different tests (Figs. 3.2–3.4). The more frequent type, called form A, is smaller in size but has a large initial chamber or **megalosphere**. The less common type, called form B, is larger in size and has a small initial chamber or microsphere.

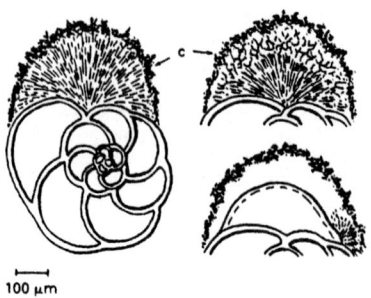

Fig. 3.2 Chamber construction in *Discorbis*; c = growth cyst (×40). After le Calvez (1938, fig. 25)

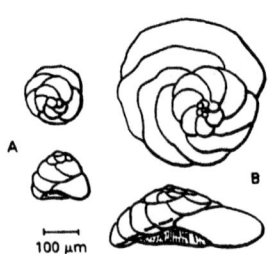

Fig. 3.3 Dimorphism in *Discorbis*: spirals seen from above and side view of megalospheric (A) and microspheric (B) individuals (×65). From le Calvez (1950, fig. 2)

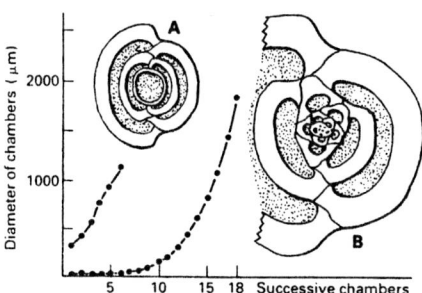

Fig. 3.4 Growth of chambers in *Pyrgo* as a function of their order of appearance: A, Megalospheric individual at first biloculine; B, Microspheric individual, successively quinqueloculine, triloculine and biloculine (×20). After Rhumbler (1959, fig. 7)

The morphological dimorphism of foraminifera is explicable in terms of a reproductive cycle in which, interposed between two periods of growth, there are separate generations, one asexual (**schizont**) and the other sexual (**gamont**). In *Elphidium crispum*, as in most species, the morphological dimorphism is superimposed on biological dimorphism:

- form A, which is megalospheric, is a uninucleate haploid gamont;
- form B, which is microspheric, is a plurinucleate diploid schizont..

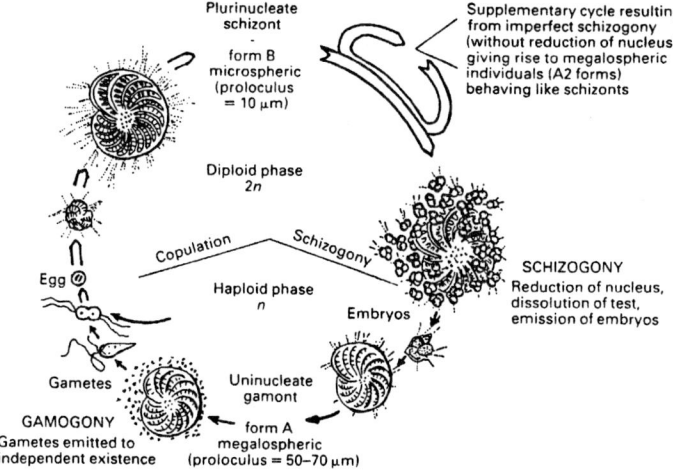

Fig. 3.5 Life cycle of *Elphidium crispum*: duration two years in the English Channel, 1 year in tropical waters. From various authors

The microspheric stage is 'conservative' in the sense that its ontogenetic development recapitulates the phylogenetic development of the group. The megalospheric stage, on the other hand, is 'progressive' in that it skips the primitive phases (tachygenesis) and starts immediately from an advanced state of evolution (Figs. 3.4 and 3.5).

Among living populations of foraminifera, the schizonts are often relatively few in relation to the gamonts, indicating that sexual reproduction is the exception. To this phenomenon must be added the destruction of microspheric tests at the moment embryos are emitted. These two reasons explain the paucity of microspheric individuals in fossil associations.

During periods of growth that last four to six times longer for schizonts than for gamonts, foraminifera feed on particles and on microscopic prey (diatoms) both living and dead. New chambers are added periodically while the creature remains within the shelter of a growth cyst formed from particles assembled by the pseudopodia (see Fig. 3.2).

Ecology

Numerous foraminifera inhabit the benthic environment. Some move freely over the sea-bed or in the first few millimetres of sediment. Others use their pseudopodia or calcareous secretions to attach themselves to supports such as rocks, shells and seaweed. Most are marine and stenohaline (they can tolerate only very small variations in the salinity of the water). Certain groups, however, having a porcelaneous test (e.g. the miliolines *Spirolina*, *Peneroplis* and *Alveolinella*) can live equally well in hyperhaline environments (lagoons with a salinity >35 parts per mille ($^o/_{oo}$)). Certain types such as the agglutinates (e.g. *Eggerella*) and hyalines (e.g. *Nonion*) prefer water with a low salinity e.g. brackish lagoons and estuaries. Still others (e.g. *Trochammina* and *Elphidium*) can adjust to considerable variations in salinity and may be found in all environments with the exception of lakes where foraminifera never live.

Foraminifera are used to interpret past water depths, since depth- and space-related parameters are of greatest significance, the foraminifera occupying different levels according to local values for temperature, oxygen content,

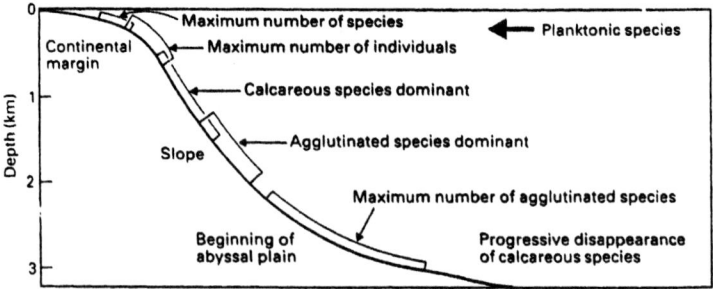

Fig. 3.6. Depth distribution of recent benthic foraminifera. After Bolstovsky and Wright (1976, fig. 55)

light, etc. As a general rule, species with a porcelaneous test live in shallow waters, whereas those with a hyaline test occur everywhere but in the deepest areas. Species with agglutinated tests are similarly ubiquitous but they alone survive at depths below 4000 to 5000 m (Fig. 3.6).

Biocoenose are more abundant and diverse in warm waters and the larger foraminifera normally occur in tropical environments ($T \geqslant 18°C$ and even 22°C) at depths not exceeding 200 m (Fig. 3.7).

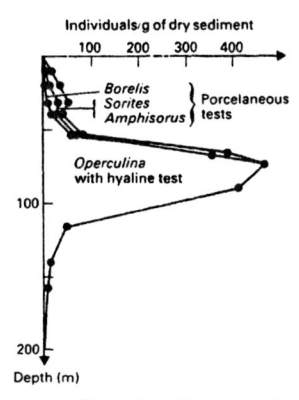

Fig. 3.7. Present depth distribution of larger benthic foraminifera in the Red Sea. After Hottinger (1977, fig. 12)

Some 40 species are planktonic and stenohaline (salinity between 34 and 36‰). The most favourable depths lie between 6 and 30 m; below 200 m they are virtually absent. The species form recognisable biogeographic associations with areas of distribution running more or less parallel to the equator.

2. THE TEST AND ITS FOSSILIZATION

Studies have been made in the minutest detail of foraminiferan tests as these fossilize very well. Under certain conditions, the basal chitinoid layer is also preserved.

Composition and Microstructure of the Wall

As far as living foraminifera are concerned, the tests are composed of:

- Either debris from the ambient environment bonded by a cement secreted by the creature; such tests are said to be **agglutinated**.
- Or calcareous, calcitic or, more rarely, aragonitic material; such tests are secreted entirely by the animal and, depending on the arrangement of the crystals, may be distinguished as **porcelaneous** or **hyaline**.

In **agglutinated tests** (Fig. 3.8), the material borrowed from the habitat is variable in nature: grains of quartz (**arenaceous tests**), flakes of mica,

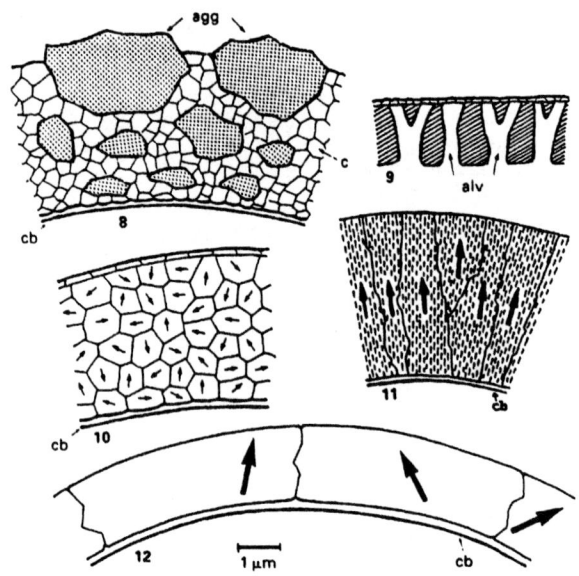

Fig. 3.8–12 Principal microstructural types of foraminiferid tests
8 Agglutinated test with compact wall: agg = agglutinated; c = cement; cb = chitinoid basal layer
9 Agglutinated test with alveolar wall: ramified and unramified alveoli opening towards the interior of the test
10-12 Schematic comparison of optical axis orientation in calcareous tests:
Porcelaneous test (10); Fibroradial hyaline test (perforations not shown) (11) and Granular hyaline test (perforations not shown) (12); arrows indicate the optical axes of the crystalline elements; cb = chitinoid basal layer Partly after Bellemo (1974, Fig. 2)

argillaceous particles, various skeletal remains such as small foraminifera, sponge spicules and coccoliths. This material may be selected by foraminifera according its size, nature and colour and is arranged in different ways in the test. The cement that is secreted is chitinoid (*Trochammina*) or calcitic; in the latter case it is often coloured with oxides of iron. The relative proportions of agglutinate and cement vary according to the species, the individual and the part of the test. The extremely thick walls of certain Lituolidae are honeycombed with **alveoli** (Fig. 3.9).

The crystals forming the wall of **porcelaneous** ('**imperforate**') tests (Fig. 3.10) are very small (0.1 to 2 μm), globular or acicular and are arranged randomly. As the crystals are abundant and as there is no parallelism between their optical axes, rays of light cannot pass through the wall. In transmitted light (or rather non-transmitted light) this appears black and opaque whereas its reflection is white and brilliant. Although the wall is sometimes pierced by cavities or pseudopores, these rarely pass all the way through. The porcelaneous tests of living foraminifera and of certain fossils have an amber tint that seems to be linked with the presence of organic matter which normally disappears in the course of fossilization.

Hyaline ('**perforate**') tests are characterized by their glass-like transparency which results:

- For the **granular** tests (Fig. 3.12), from the thinness of the walls which are formed from the juxtaposition of a limited number of microcrystals (5 to 10 μm) with variable optical orientation.
- For the **fibroradial** tests (Fig. 3.11), from the regular arrangement of fine acicular crystals with an optical axis perpendicular to the surface of the test; should the wall grow thicker, the transparency may be reduced but it can still be detected in thin sections.

All hyaline tests are traversed by perforations with variable diameter (0.5 to 15 μm), density and location.

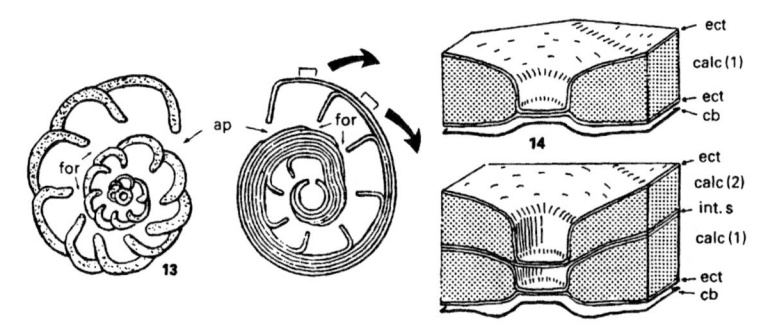

Fig. 3.13 Non-lamellar test (left) and lamellar test (right): ap = aperture; for = foramen. After Loeblich & Tappan (1964, fig. 54) and Reiss (1957, fig. 2b)

Fig. 3.14 Wall structure of a lamellar hyaline test at a perforation: a monolamellar wall (above); a lamellar wall (below); ect = ectoplasm; int. s = interlamellar ectoplasmic sheet; calc (1,2) = primary (secondary) calcareous layer; cb = chitinoid basal layer. After Sliter (1974, fig. 1)

Contrary to the preceding examples, many hyaline tests are **lamellar** (figs 3.13 and 14). They have walls formed from the superposition of calcareous sheets separated by proteinaceous layers. The thicker the wall, the older the chamber. The reason for this is that whenever a new chamber is added, the foraminifera covers the free surface of all the older chambers with a calcareous sheet. Finally, the hyaline tests of diverse species (including the Nummulitidae) are traversed by a more or less complex system of **channels**.

The tests of Palaeozoic foraminifera are unlike those of living species as their walls display particular microstructures described as **microgranular, diaphanothecate, keriothecate** or **pseudofibrous** which are still imperfectly understood. Detailed examples of these are given below.

Morphology of Chambers

Chambers are spherical to flattened in shape. They are separated from each other by partitions (or septa) the sutures of which are rectilinear, curved or winding.

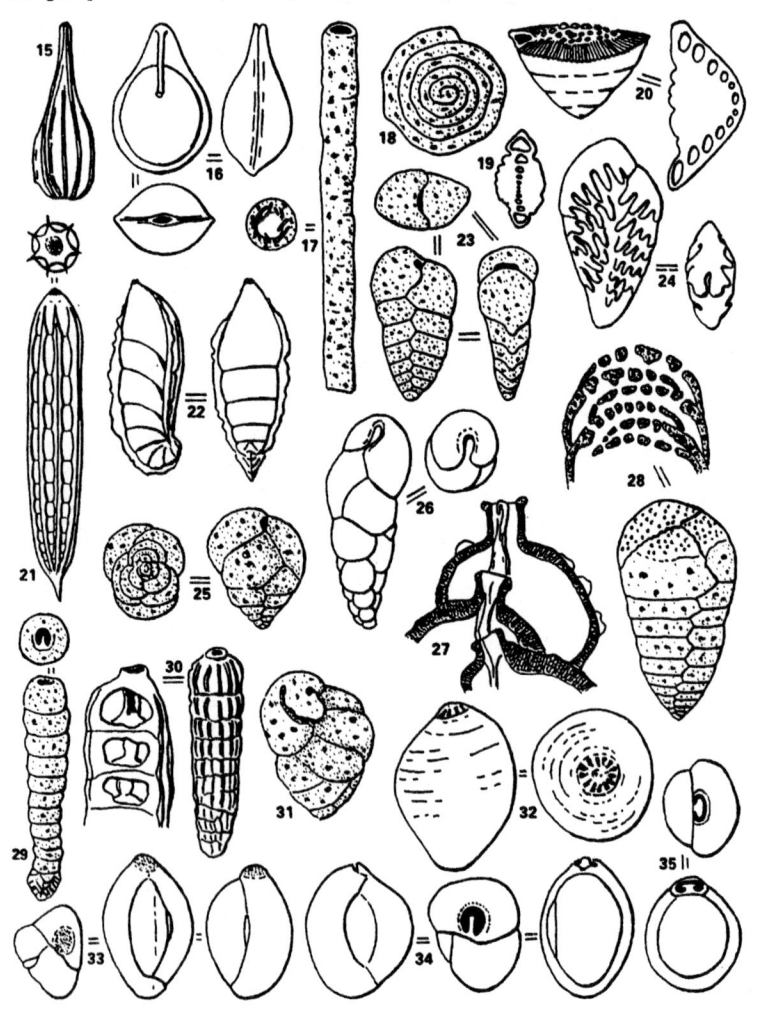

The initial chamber is called the **proloculus**. Sometimes the first chambers are grouped in an **embryont** or **juvenarium**. the postembryonical chambers are of increasing size. The last one communicates with the exterior through an **aperture**, which is enormously variable in form and position.

There may be no aperture at all. It may be simple, in this case taking the shape of a circle, a slit or a crescent; or it may be radial or dendritic, sometimes at the top of a neck or partially covered by calcareous projections (**tooth, valve, trematophore** or **tegillum**). There may also be multiple apertures with several small orifices arranged in a row (**linear aperture**) or randomly (**cribrate aperture**). Its position in relation to the chamber may be basal, terminal, sutural or peripheral. If the test is a trochospiral, the position may be median, umbilical or spiral. Apart from the opening of the last chamber (the **principal aperture**), there may also remain **supplementary** or **relict apertures** of preceding chambers. Foramina which interconnect the chambers represent earlier apertures or result from the secondary reabsorption of a part of the partition. Rows of foramina are called **stolons**.

Figs 3.15–35 Morphology of foraminifera with monolocular tests (15–20), rectilinear unilocular tests (21–31), porcelaneous bunched tests (32–35)
15 *Lagena* (Jurassic–Recent): hyaline, oral neck, striated ($\times 55$)
16 *Fissurina* (Cretaceous–Recent): hyaline, internal oral tube ($\times 100$)
17 *Bathysiphon* (Cambrian ? or Ordovician–Recent): agglutinated, tubular, rectilinear; left, radial axis ($\times 8$)
18 *Ammodiscus* (Silurian–Recent): agglutinated, tubular, planispiral ($\times 60$)
19 *Involutina* (Triassic–Lias): calcitic, tubular, planispiral; axial section ($\times 12$)
20 *Trocholina* (Triassic–Cretaceous): calcitic, tubular, trochospiral; lateral view (left) and axial section (right) ($\times 30$)
21 *Nodosaria* (Permian ? or Jurassic–Recent): hyaline, uniserial, ribbed, radial terminal aperture ($\times 10$)
22 *Saracenaria* (Jurassic–Recent): hyaline, uniserial beginning planispiral, keeled, radial terminal aperture ($\times 50$)
23 *Textularia* (Jurassic–Recent): agglutinated, biserial, basal aperture ($\times 30$)
24 *Bolivinoides* (Cretaceous–Palaeocene): hyaline, biserial, digitate sutures, basal aperture ($\times 50$)
25 *Eggerella* (Cretaceous–Recent): agglutinated, triserial beginning multiserial, basal aperture ($\times 30$)
26 *Bulimina* (Palaeocene–Recent): hyaline, triserial, basal aperture ($\times 50$)
27 *Uvigerina* (Eocene-Recent): hyaline, triserial, terminal aperture at the top of the neck; dissection showing the internal oral tube ($\times 100$)
28 *Chrysalidina* (Cretaceous): agglutinated, triserial, sieve-plate terminal aperture; above, axial section of last chambers ($\times 10$)
29 *Clavulina* (Palaeocene–Recent): agglutinated, uniserial beginning triserial, terminal aperture with tooth ($\times 25$)
30 *Siphogenerinoides* (Cretaceous–Palaeocene): hyaline, uniserial beginning biserial, ribbed, aperture at the top of a neck ($\times 40$); dissection showing the internal oral tube ($\times 100$)
31 *Valvulina* (Triassic–Recent): agglutinated, triserial beginning multiserial, basal aperture with valve ($\times 25$)
32 *Idalina* (Cretaceous–Eocene): unilocular, trematophore terminal aperture ($\times 6$)
33 *Miliola* (Eocene): quinquelocular, trematophore terminal aperture ($\times 20$)
34 *Triloculina* (Jurassic–Recent): trilocular, toothed terminal aperture ($\times 30$)
35 *Pyrgo* (Jurassic–Recent): bilocular, toothed terminal aperture ($\times 20$)

After Barr, Brady, Cushman, Kaaschieter and le Calvez.

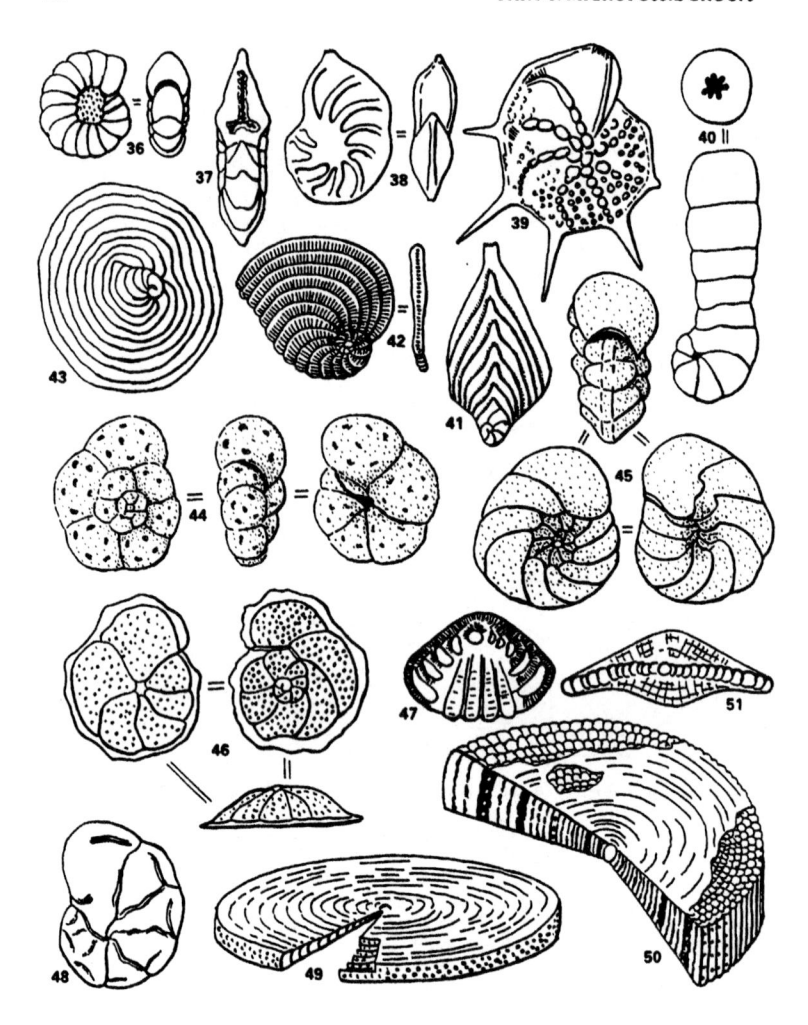

Figs. 3.36–51 Morphology of foraminifera with multilocular tests. Planispiral (36–41), trochospiral (42–48), annular (49–51).

36 *Nonion* (Palaeocene–Recent): hyaline, umbilical granules, basal aperture (×60)
37 *Dendritina* (Eocene–Recent): porcelaneous, dendritic aperture (×30)
38 *Lenticulina* (Triassic–Recent): hyaline, limbate sutures, radial terminal aperture (×20)
39 *Lenticulina* (Triassic–Recent): granular sutures, peripheral spines (×15)

The interior of the chamber may be empty (simple) or contain tubes or tooth plates linking the foramina. In the more complex forms (the 'larger' foraminifera), the chambers are partially occupied by skeletal formations incorrectly called endoskeletons. These internal structures consist of **tectoria, parachomata** and **chomata** (in the Fusulinidae), **pillars, septulae,** etc.

Arrangement of Chambers

The simplest tests are **unilocular**, the single chamber being spherical or tubular. More often, however, tests are **multilocular**. Some examples will serve to emphasize the extreme variety of test architecture (Figs. 3.15–3.51).

In rectilinear systems, the chambers are arranged on a straight or curved axis. If there is only one series of chambers, the test is **uniserial**. If there are several, it may be **biserial, triserial,** or **multiserial**.

When the growth axis is helical, the chambers are arranged in a spiral. If this is a plane coil, the **planispiral** test has two identical faces. The coiling is said to be **evolute** if all the whorls of the spiral are visible laterally; it is **involute** when the last whorl covers all the preceding ones. The point at which sutures converge is the **umbilicus**.

If the spiral is trochoid, the **trochospiral** test has opposite sides that are different: an evolute or spiral side and an involute or umbilical side. The direction taken by a spiral is not always uniform for all the individuals of the same species: depending on the climatic conditions, it may be right-handed or left-handed. **Annular** or **cyclic** tests result when a spiral begins but is enveloped in concentric chambers.

The Miliolidae have a **streptospiral** arrangement (Figs 3.32–35). The arched

---

40 *Spirolina* (Cretaceous–Recent): porcelaneous, planispiral then uniserial rectilinear, dendritic terminal aperture (×50)
41 *Neoflabellina* (Cretaceous–Palaeocene): hyaline, planispiral then uniserial rectilinear with chambers in herringbone pattern, radial terminal aperture (×25)
42 *Peneroplis* (Eocene–Recent): porcelaneous, last chambers embracing, multiple peripheral apertures (×20)
43 *Archiacina* (Oligocene) = *Peneroplis* with annular final chambers (×20)
44 *Trochammina* (Carboniferous–Recent): agglutinated, umbilical basal aperture with lip; from left to right, views of spiral, profile and umbilicus (×40)
45 *Gavelinella* (Cretaceous–Eocene): hyaline, peripheral-umbilical basal aperture; from left to right, views of spiral, profile and umbilicus (×60)
46 *Cibicides* (Cretaceous–Recent): hyaline, spiral basal aperture, plane spiral face (attachment); from left to right, views of umbilicus, spiral and profile (×60)
47 *Lockhartia* (Palaeocene–Eocene): hyaline, umbilical granules; axial section (×10)
48 *Epistomaria* (Eocene): hyaline, umbilical basal principal aperture, supplementary sutural and peripheral apertures; umbilical view (×30)
49 *Cyclolina* (Cretaceous): agglutinated, concentric chambers, multiple peripheral apertures (×30).
50 *Orbitolites* (Eocene): porcelaneous, concentric series of arcuate chambers, multiple peripheral apertures (×8)
51 *Linderina* (Eocene): hyaline, concentric series of arcuate chambers flanked by lateral thickenings; axial section (×50)

After Guyader, le Calvez, Loeblich & Tappan and Marie.

chambers, tangential at their two extremities with the extension axis, are arranged in cycles of five, three or two loculi or one loculus. Each new chamber has its aperture facing the aperture of the preceding chamber.

In addition to these simple modes, tests may exhibit several successive modes of arrangement. In such cases, the mode of arrangement is said to be mixed or composite. Finally, the orbitoid mode is characterized by the presence of two sets of different chambers. The plane of the cyclical equatorial chambers is bordered on either side by successive arched layers of lateral chamberlets (Figs. 3.103 and 104).

Form and Orientation of Tests

The general form and dimensions of tests depend, to a large extent, on the form, number and arrangement of chambers. Account must also be taken of ornamentation (i.e. any marked calcareous external protrusions):

- Surface of the test: is it smooth or covered with striations, ribs, tubercules, spines, a reticulum, etc?
- Sutures: are they simple or limbate (i.e. underlined by a band of variable width)?
- Periphery of the test: is it accentuated or not by spines, keels?
- Umbilicus: is it vacant or occupied by one or more granules?

In the main, the average size of tests varies between 0.1 and 1 mm. The largest (nummulitids), however, can be over 10 cm.

With the exception of conical forms, it has become the custom to illustrate tests with their aperture at the top. Thin sections are oriented in relation to an axis: the growth axis for rectilinear forms and the spiral axis for coiled and streptospiral forms. The most representative sections are axial (passing through one of these axes) and transverse (perpendicular to them).

3. SYSTEMATIC SURVEY

With 1400 genera and 30 000 species (including 4500 still in existence), the order Foraminiferida accounts for half the known Protozoa (i.e. 2.5% of all organisms described to date). Their classification is difficult and is entirely

---

Figs 3.52–62 Large agglutinated foraminifera: Lituolidae (52,53), Orbitolinidae (54–62). Arrows indicate the direction of growth of tests; ap = aperture; for = foramen; sut = suture

52 *Pseudocyclammina* (Lias–Cretaceous): external views (left) (×30) and transverse section (right) (×80) showing the alveolar wall; alv = alveolus

53 *Labyrinthina* (Triassic–Jurassic): dissection showing the periphery of the chambers occupied by radial septula (×20)

54–56 *Fallotella* = *Coskinolina* auct. (Palaeocene): 54, external view (×30); 55, transverse sections (above); axial sections (below) (×20); 56, diagram of fragments of two successive chambers (×70); sep = septum; rsep 1 and 2 = primary and secondary peripheral radial septula; cap = cribrate aperture; pil = central pillar

57 *Dictyoconus* (Eocene): above, axial sections (×12); below, transverse sections

FORAMINIFERA

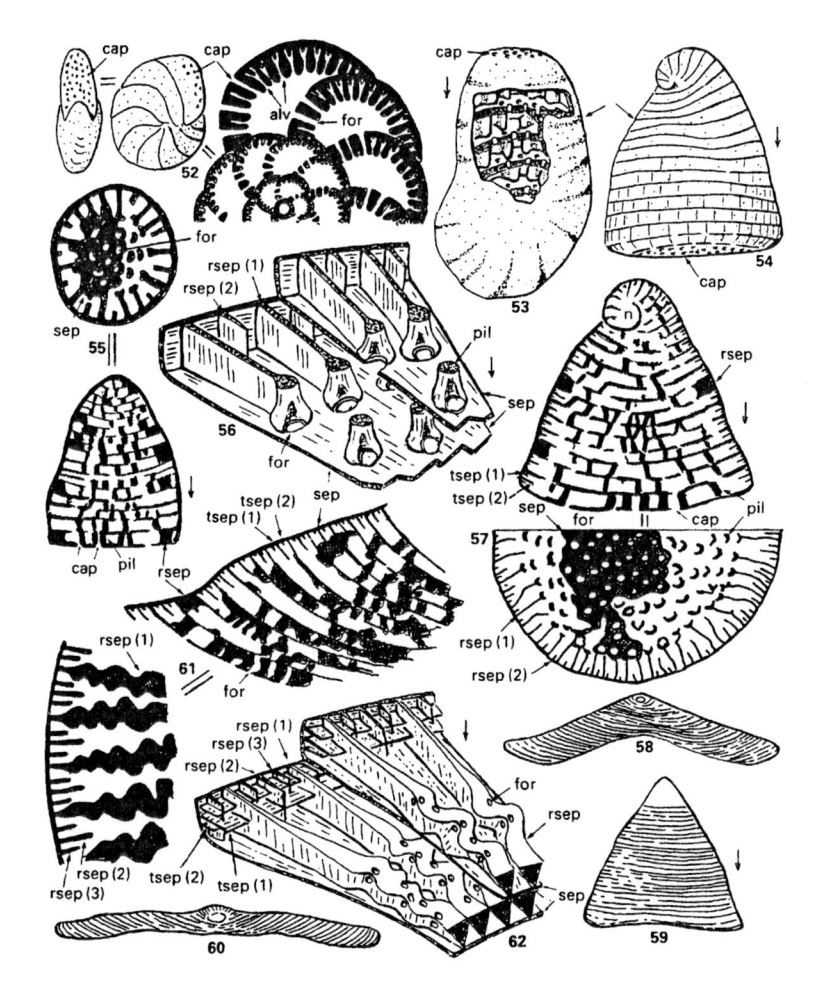

(×40); same abbreviations as for preceding figures with the addition of tsep 1 and 2 = primary and secondary transverse septula, and n = nucleoconch
58-62 *Orbitolina* (Cretaceous): 58, *O. texana* (Roemer, 1849) axial section: (×10); 59, *O. conulus* Douville, 1912: axial section (×20); 60, *O. concava* (Lamarck, 1801): axial section (×3); 61, portion of axial section (above) (×30); portion of transverse section (below) (×70); 62, diagram of fragments of two successive chambers. Diagrams of orbitoline nucleoconches are given in Fig. 13.15

After Banner, Cati, Davies, and Douglass and from original drawings.

based on the characteristics of the test. The most recent attempt has been by Loeblich and Tappan (1964). Although modified, it remains very artificial. Considerable amendments are likely to follow once there is a better knowledge of the biology of modern forms and of the composition of the proteins contained in the tests.

Traditionally, a contrast has been made between the 'small' and the 'large' foraminifera: the former, taken from screening residues, are determinable only from their external characteristics; the latter, with a complex internal structure, are recognizable only after examination in thin sections.

Textulariina

The suborder Textulariina encompasses the genera with agglutinated tests: the Ammodiscidae, unilocular, globular, tubular (Fig. 3.17) or spiral (Fig. 3.18), and forms that are multilocular and rectilinear (Figs 3.23, 25, 28, 29 and 31) or at least initially spiral. Among the latter, the Lituolidae are of particular interest (Figs 3.49 and 52, 53). These are planispiral but become rectilinear or annular. They have a wall that is sometimes alveolar and chambers may be vacant or occupied by pillars and septula. The Orbitolinidae (Figs 3.54–62) start out briefly with a trochospiral, which is followed by a rectilinear conical stage with low chambers. These grow rapidly in diameter and are occupied by septula on the periphery and by primary radial septula in the centre.

Fusulinina

To the suborder Fusulinina belong the Palaeozoic foraminifera with a structure

Fig. 3.63–77 Palaeozoic foraminifera, Parathuramminidae, Endothyridae (63–70) and Fusulinidae (71–77).
63 Diagram of the wall of an endothyrid: microgranular external wall and pseudofibrous internal layer
64 *Earlandia* (Devonian–Carboniferous): microgranular, unilocular, tubular; from left to right, transverse, oblique and axial sections ($\times 30$)
65 *Climacammina* (Carboniferous–Permian): two-layered wall, biserial then uniserial, cribrate terminal aperture; axial section ($\times 10$)
66 *Geinitzina* (Devonian–Permian): pseudofibrous, uniserial, simple terminal aperture; axial section ($\times 80$)
67 *Archaediscus* (Carboniferous): two-layered wall, monolocular, tubular, pseudoplanispiral; axial section ($\times 100$)
68 *Tetrataxis* (Carboniferous–Permian); two-layered wall, multilocular, trochospiral; from top to bottom, external views of profile, spiral and axial section ($\times 15$)
69 *Carbonella* (early Carboniferous): microgranular, planispiral, early stage tubular, later septate; transverse section ($\times 30$)
70 *Endothyra* (Carboniferous–Permian): microgranular, planispiral with an irregular early stage; axial (left) and transversal (right) sections ($\times 25$)
71 *Dunbarula* (Permian): portion of transverse section showing wall structure with external tectorium (et), tectum (t) diaphanotheca (d) and internal tectorium (it) ($\times 100$)
72 Keriothecae wall with perforate tectum (t), internal keriothecae (ik) and external keriothecae (ek).
73 *Ozawainella* (Carboniferous–Permian): tectum, chomata (in black) on either side of the foramina ($\times 40$)
74 *Staffella* (Carboniferous–Permian): tectum and diaphanotheca, rectilinear septa, chomata ($\times 10$)

FORAMINIFERA 33

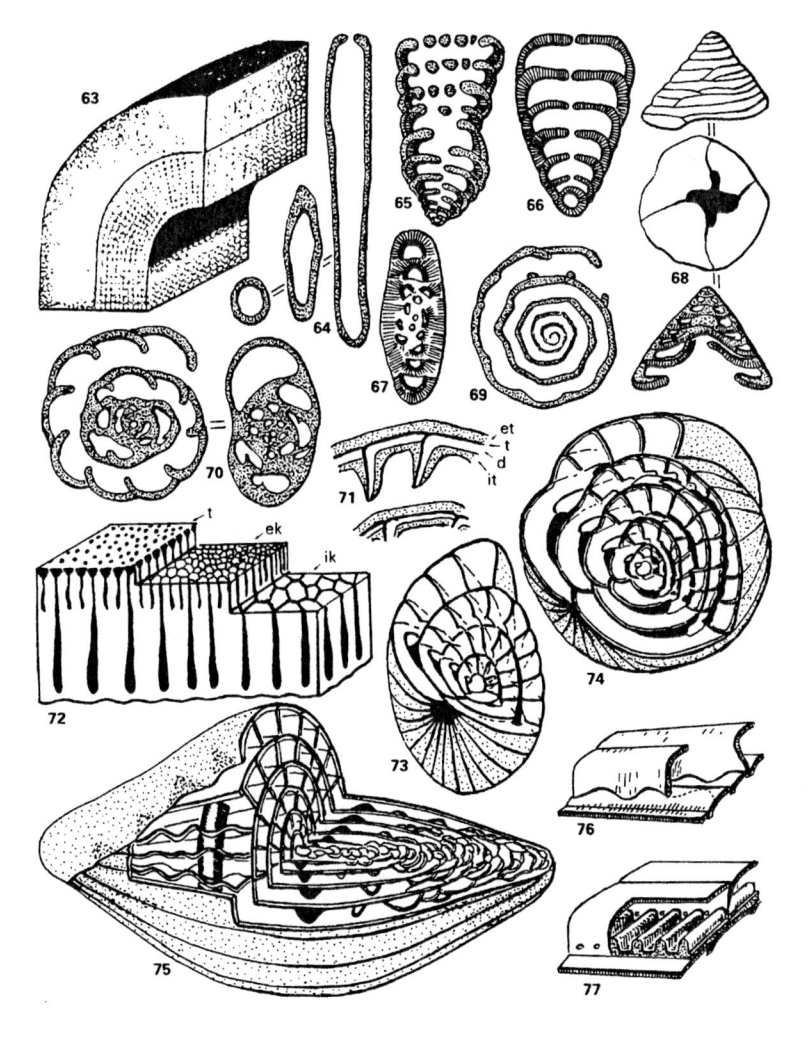

75 *Fusulinella* (Carboniferous–Permian): tectum and diaphanotheca, septa folded at poles, chomata (×20)
76 Folded septa separating empty chambers
77 Rectilinear septa separating chambers occupied by a basal 'endoskeleton' or parachomata

After Ciry, Cummings, Dain, Rauzer-Chernousova, and Ross, etc.

that is microgranular and/or pseudofibrous; Parathuramminidae–Endothyridae (Figs 3.63–70) with simple chambers that may be single or rectilinear or variously spiral; and the Fusulinidae (Figs 3.71–80) with a robust test that may be planispiral, lenticular or fusiform. About 100 genera can be distinguished on the basis of the microstructure of the wall (**diaphanothecate** or **keriothecate**), the appearance of the partitions (rectilinear or folded) and the characteristics of the 'endoskeleton'.

Miliolina

The suborder Miliolina or species with porcelaneous tests include the streptospiral Miliolidae (Figs 3.4, and 32–35), the Soritidae, which are initially planispiral (Figs 3.37, 40, 42, 43 and 50) and the Alveolinidae (Figs. 3.81–86) which are planispiral, fusiform, and contain chambers divided by septula parallel to the direction of the spiral.

Rotaliina

The classification of the suborder Rotaliina, the foraminifera with hyaline tests, is far from consistent. First of all, it must be noted that there are several aberrant genera with a monocrystalline aragonitic or calcitic test. Putting it as simply as possible, the true hyalines fall into four groups:

- Small forms without canal systems such as the planispiral or rectilinear Nodosariidae with radial aperture (Figs 3.15, 21, 22, 38, 39 and 41); the Buliminidae, which are multiserial and have a slit aperture and an internal oral appendage (Figs 3.16, 24, 26, 27 and 30); and the trochospiral Discorbidae (Figs 3.2, 3, 45, 46, 48 and 51).
- Genera with a canal system: the trochospiral Rotaliidae and the large planispiral Nummulitidae (Fig. 3.87–100).
- The Orbitoididae (Fig. 3.103–108), which, though without a canal system, are none the less the most complex of foraminifera with their multilocular juvenarium and their two types of chamber: **equatorial** and **lateral**. The Miogypsinidae (Fig. 3.101 and 102) with dissymmetric test, eccentric initial chambers and planispiral form are sometimes classified alongside the lenticular and symmetric Orbitoididae.
- The planktonic foraminifera forming the fairly homogeneous group of Globigerinidae (Fig. 3.119–137). They have spherical chambers, few in number, are sometimes keeled, and are always ornamented with thin spines that do not withstand fossilization and support a bubbly ectoplasm in life.

Fig. 3.78–86 Larger foraminifera: Fusulinidae (contd) (78–80) and Alveolinidae (81–86);

78 *Pseudofusulina* (Carboniferous–Permian): tectum and keriotheca, folded septa, axial filling (in black) (×6)

79 *Lepidolina* (Permian): above, portions of transverse sections with visible septa (sep); below, axial sections; t = tectum and keriotheca; the 'endoskeleton' is mural (or keriothecate) and basal (parachomata) giving incomplete axial septula (asep) and complete transverse septula (tsep)

80 *Neoschwagerina* (Permian): tectum and keriotheca, rectilinear septa, 'endoskeleton' that is mural (or keriothecate) and basal (parachomata) (×55)

81 *Ovalveolina* (Cretaceous): continuous transverse septula (×30)

FORAMINIFERA 35

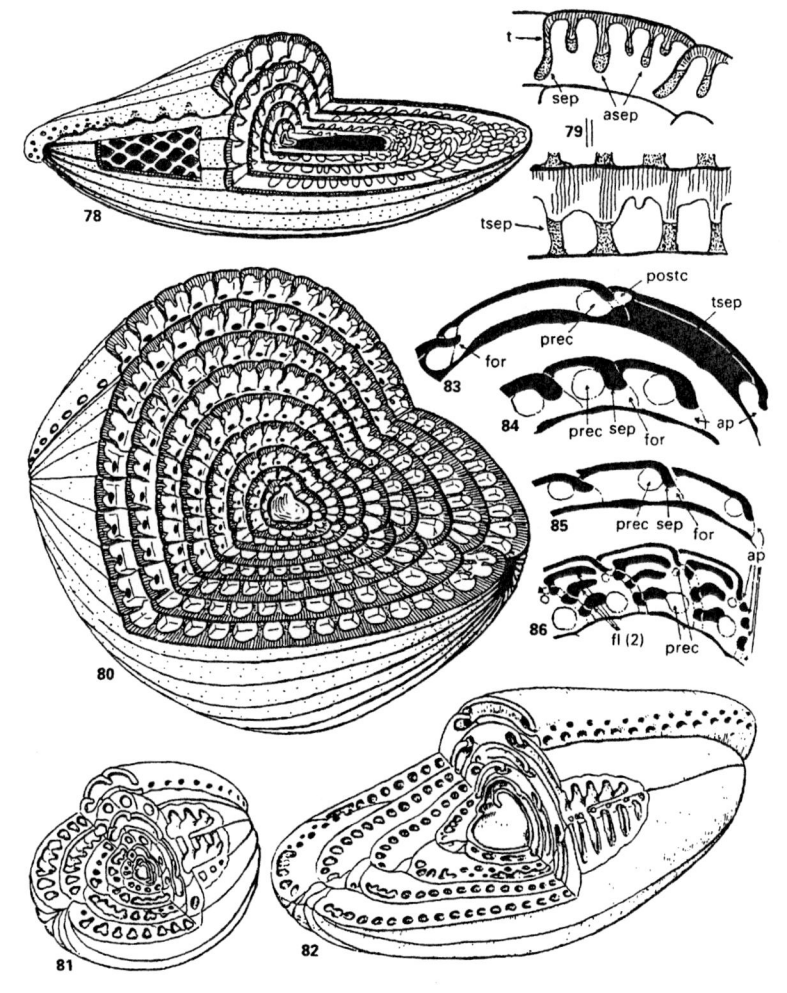

82 *Alveolina* (Eocene): alternate transverse septula (×30)
83–86 Portions of transverse sections: 83, *Alveolina;* 84, *Ovalveolina;* 85, *Praealveolina* (Cretaceous); 86, *Alveolinella* (Miocene–Recent); ap = aperture; for = foramen; sep = septum; tsep = transverse septulum; prec = preseptal canal; postc = postseptal canal; fl(2) = secondary floor (×65). For axial sections of *Alveolina*, see Fig. 13.20.

After Dutkevitch, Miklukho-Maklai, Reichel and Skinner.

Some types have rectilinear tests that are either biserial or planispiral. Most, however, are trochospiral and some have a final enveloping chamber (*Orbulina*, Fig. 13.18).

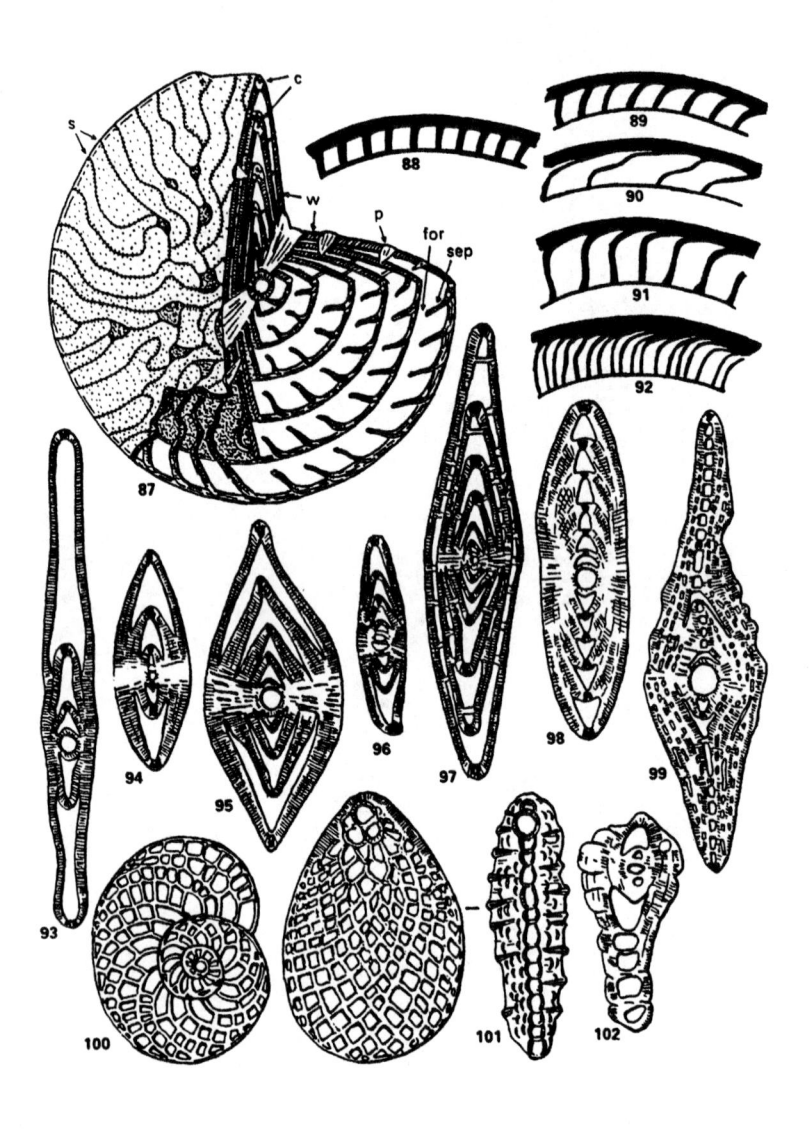

4. FORAMINIFERA THROUGHOUT GEOLOGICAL TIME

The foraminifera appeared in the Ordovician and perhaps even in the Cambrian (the ammodiscid *Bathysiphon*), as forms with a unilocular agglutinated test.

Microgranular calcareous tests emerged in the Silurian and pseudofibrous types in the Devonian. An important event took place at the Devonian – Carboniferous boundary: this was the development of partitions (*Carbonella*, Fig. 3.69) yielding multilocular tests with periodic growth. Not long afterwards, during the Carboniferous, porcelaneous tests appeared together with morphological dimorphism and trochospiral coiling: these occurred in the agglutinated *Trochammina* (Fig. 3.44) and the calcareous *Tetrataxis* (Fig. 3.68). The Endothyridae and the Fusulinidae proliferated in the late Palaeozoic before disappearing on the threshold of the Mesozoic.

Although poor in foraminifera, the Triassic saw a considerable renaissance which, during the Jurassic, took the form of the development of the true Miliolidae, the first hyaline tests to be fibroradial, perforate and lamellar (Nodosariidae), and planktonic forms (*Globuligerina*).

During the early Cretaceous, some species began to populate lagoons. Diversification continued in the Cretaceous with the acquisition of a canal system (Rotaliidae), the proliferation of planktonic forms (*Globotruncana*) and the larger benthic groups (Orbitolinidae, Alveolinidae and Orbitoididae).

The beginning of the Cenozoic witnessed the sudden extinction of many genera both benthic (*Orbitoides*) and planktonic (*Globotruncana*). This was soon followed by a new and important phase of diversification affecting old

---

Fig. 3.87–102 Larger hyaline foraminifera: Nummulitidae (87–100) and Miogypsinidae (101,102).

87 *Nummulites* (Eocene–Oligocene–Recent): involute test; c = marginal cord; for = foramen; lc = last chamber; p = pillar or surface granule (×6); s = suture with transverse trabeculae; sep = septum; w = perforate wall

88–92 Equatorial transverse sections showing septa in: 88, *N. discorbinus* (von Schlotheim, 1820); 89, *N. striatus* (Bruguiere, 1792); 90, *N. aturicus* (Joly & Leymerie, 1848); 91, *N. bolcensis* Munier-Chalmas, 1877; 92, *N. murchisoni* Rutimeyer, 1850

93,94 *Operculina* (Palaeocene–Recent): involute–evolute test; axial sections of: 93, *O. complanata* (Defrance, 1822) and 94, *O. ammonea* Leymerie, 1846) (×40)

95–97 *Nummulites*: involute test; axial sections of: 95, *N. chavanesi* De la Harpe, 1878 (×15); 96, *N. exilis* Douville, 1919 (×10); 97, *N. planulatus* (Lamarck, 1804) (×10)

98 *Assilina* (Eocene): evolute test; axial section (×6)

99 *Spiroclypeus* (Eocene–Miocene): involute–evolute test, chambers divided into chamberlets (as in the next example) but flanked by lateral chamberlets; axial section (×25)

100 *Heterostegina* (Eocene–Recent): involute–evolute test, chambers divided into chamberlets but without lateral chamberlets; transverse section (×12)

101 *Miogypsina* (Miocene): transverse sections on the left, axial sections on the right (×20)

102 *Miogypsinoides* (Oligocene–Miocene): axial section (×20)

After Arni, Bannink, Cole, Decrouez, Drooger, Hansawa, Schaub and from original drawings.

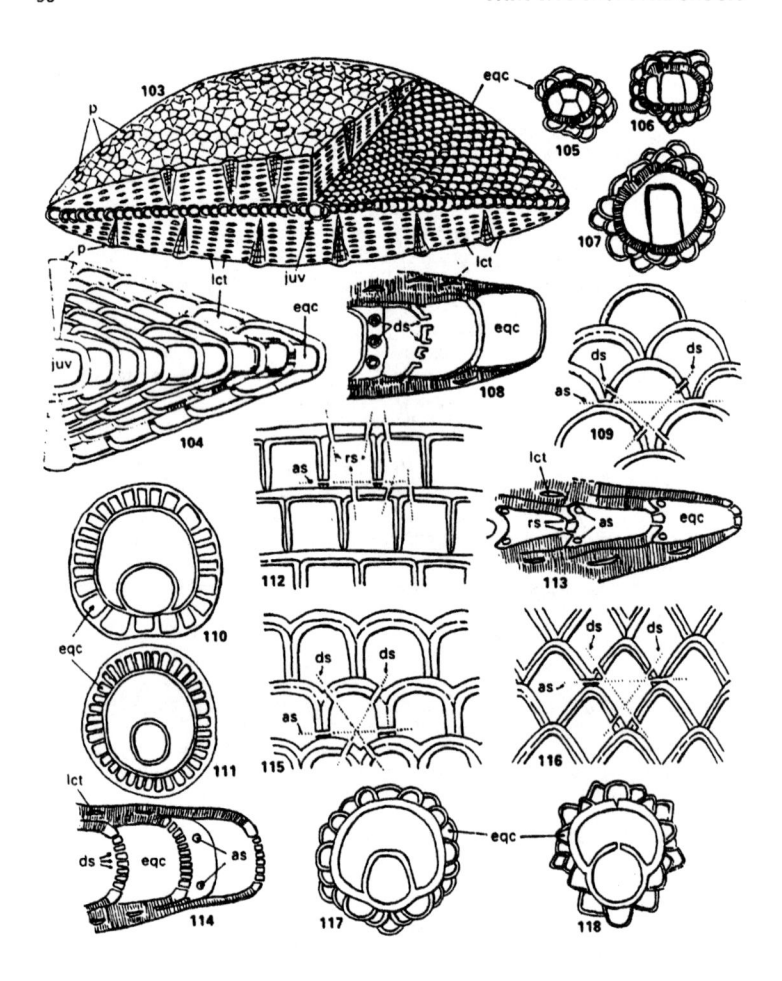

Fig. 3.103–118 Larger foraminifera (contd.): Orbitoidae

103–108 *Orbitoides* (Late Cretaceous): **103**, block diagram (×25); **104**, diagram of the axial section

105–107 Megalospheric juvenarium: **105**, quadrilocular; **106**, trilocular; **107**, bilocular; **108**, equatorial chambers in axial section (×100). juv = juvenarium; eqc = equatorial chambers; lct = lateral chamberlets; p = pillar or surface granule; ds = diagonal stolon

109 *Lepidorbitoides* (Late Cretaceous): equatorial chambers in transverse section (×75); as = annular stolon; ds = diagonal stolon

110–113 *Discocyclina* (Palaeocene–Eocene): 110, juvenarium; 111, juvenarium (×50); 112, equatorial chambers in transverse section (×300); 113, equatorial chambers in axial section (×300); as = annular stolon; rs = radial stolon
114–118 *Lepidocyclina* (Eocene–Miocene). Abbreviations as above
114 Equatorial chambers in axial section (×100)
115–116 Equatorial chambers in transverse section (×100): 115, eulepidina type; 116, nephrolepidina type
117, 118 Types of juvenarium: 117, eulepidina or "trybliolepidina"; 118, nephrolepidina

After Neumann (1967).

---

groups (Alveolinidae, Orbitoididae and planktonic forms) and engendering new ones (Nummulitidae).

The end of the Oligocene was also a critical period with a sharp decline in the Lituolidae, the Alveolinidae and the Nummulitidae. The Orbitoididae disappeared in the Miocene. In Recent times the larger foraminifera are represented by no more than a few genera but the smaller benthic and planktonic forms are still present in large numbers.

Figures 3.119–137 on p.40
Fig. 3.119–137 Mesozoic (119–128) and Cenozoic planktonic foraminifera (129–137): k = keel; u = umbilicus; t = tegilla

119 *Globuligerina* (Jurassic–early Cretaceous) (×150)
120 *Heterohelix* (Cretaceous): biserial (×55)
121 *Pseudotextularia* (late Cretaceous): biserial then multiserial (×55)
122 *Hedbergella* (Cretaceous): spherical chambers, umbilical–peripheral aperture; below, axial section (×65)
123 *Praeglobotruncana* (late Cretaceous): keel (×75)
124 *Rotalipora* (late Cretaceous): keel, supplementary umbilical and sutural apertures (×50)
125–128 *Globotruncana* (late Cretaceous): two keels, large umbilicus, umbilical apertures masked by tegillae (×50). 126, *G. contusa* (Bolli, 1945): axial section; 127, *G. lapparenti* Brotzen, 1936: axial section; 128, *G. arca* (Cushman, 1926) with broken tegillae (×40)
129 *Hantkenina* (Eocene): planispiral, peripheral spines (×35)
130 *Globoconusa* (Palaeocene): supplementary sutural apertures on spiral side (×100)
131 *Acarinina* (Palaeocene–Eocene): above, axial section (×80)
132 *Morozovella* (Palaeocene–Eocene): two robust keels, narrow umbilicus; left, axial section (×75)
133 *Globorotalia* (Eocene–Recent): keel, narrow umbilicus; left, axial section (×75)
134 *Globigerapsis* (Eocene): last chamber embracing, sutural apertures on spiral side (×65)
135 *Globoquadrina* (Oligocene–Recent): above and right, axial section (×100)
136 *Globigerina* (Eocene–Recent): umbilical aperture (×75)
137 *Sphaeroidinella* (Miocene–Recent): last whorl with embracing chambers, the projecting edge of which masks the sutural apertures; on the right, axial section (×55). *Orbulina* (Miocene–Recent): last chamber completely embracing: cf. Fig. 13.18

After Bolli, Loeblich & Tappan, Postuma, Subbotina.

40 PART 1: MICROFOSSIL GROUPS

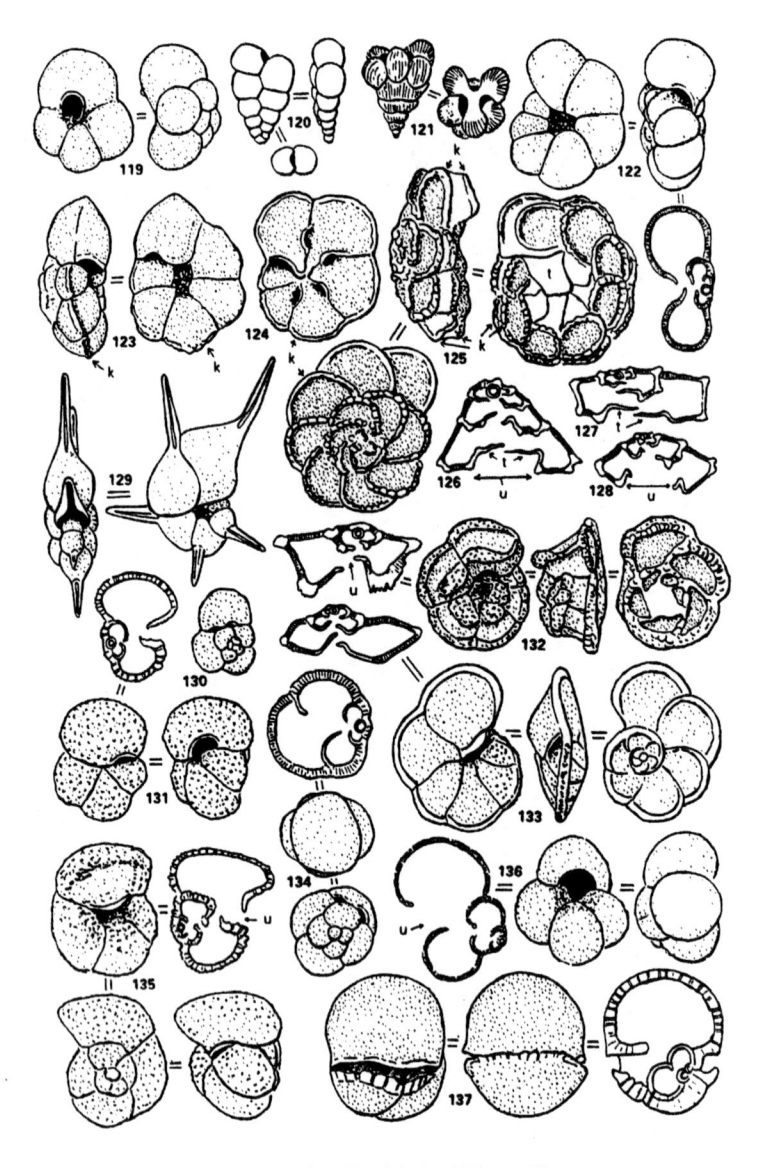

Legends to Figs 3.119 – 137 on p. 39

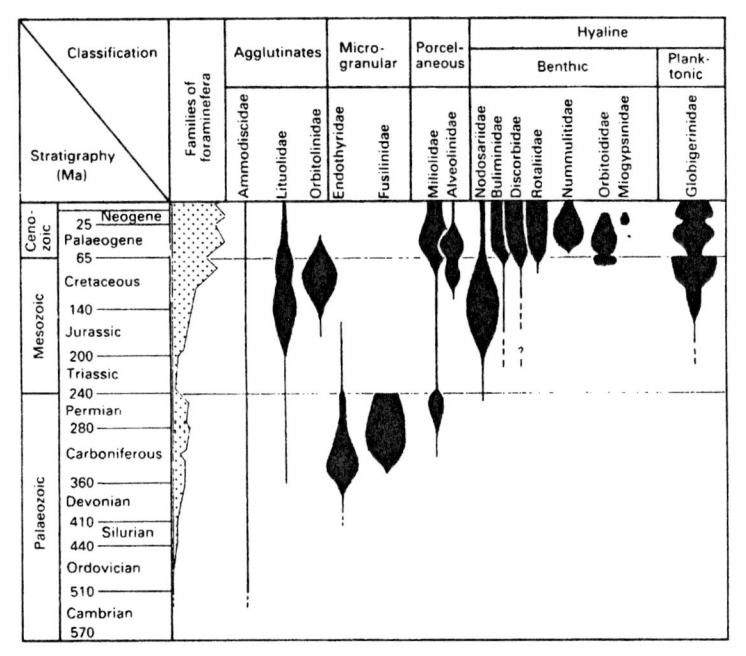

Fig. 3.138 Stratigraphic range of some foraminiferan groups

CONCLUSION

By virtue of their abundance, variety and ease of study, foraminifera could be called the pilot group that has led the way in micropalaeontology. Many methods for stratigraphical and palaeogeographical analysis of sediments were tested on foraminifera before being extended to other groups of microfossils.

BIBLIOGRAPHY

The standard work is A.R. Loeblich & H. Tappan, *Protista 2: Sarcodina, chiefly "Thecamoebians" and Foraminiferida*, in R.C. Moore (ed), *Treatise on Invertebrate Paleontology, Part C*, vols 1 and 2 (Geological Society of America and University of Kansas Press, Lawrence 1964). Other information can be gleaned from older works by J.A. Cushman, *Foraminifera: Their Classification and Economic Use*, 5th edn (Harvard University Press, Cambridge, Mass. 1959) and D.M. Rauzer-Chernousova & F.V. Fursenko, in Y.A. Orlov (ed.). [*Fundamentals of Palaeontology: Protozoa*] (Academy of Sciences, Moscow, 1959; English translation, Israel Program for Translations, Jerusalem, 1962). J.R. Haynes, *Foraminifera* (Halsted Press, New York, 433pp., 1981) and M.A. Buzas and B.K. Sen Gupta, *Foraminifera* (University of Tenessee Press, 219pp., 1982) are more recent. Another work to be noted is the incomplete but valuable M. Neumann, *Manuel de Micropaléontologie des Foraminféres*, vol 1 (Gauthier-Villars, Paris, 1967).

Monographs treating a single family or genus of foraminifera are too numerous to be cited here, but mention should be made of J.A. Postuma, *Manual of Planktonic Foraminifera* (Elsevier, Amsterdam, 1971) and G. Jenkins & J.W. Murray, *Stratigraphical Atlas of Fossil Foraminifera* (Ellis Horwood, 1981); this latter book gives a good idea of the associations found in northwestern Europe.

The prime source of information is the voluminous publication (more than 40 000 fiches) by B.F. Ellis & A.R. Messina, *Catalogue of Foraminifera* (American Museum of Natural History, New York, 1940–). If this is unavailable, the more significant species can be determined from the *Catalogue of Index Foraminifera* (3 vols, 1965–1967) and the *Catalogue of Index Smaller Foraminifera* – (3 vols, 1968–1969) published by the same authors.

Chapter 4

Ostracods

Crustacean arthropods are variously represented in washing residues and thin sections by:
- Cirripeds.
- Conchostraceous branchiopods (see Fig. 4.8). These are fairly frequent in lake and lagoon environments of the Carboniferous and the Triassic. They have a bivalved carapace, a low degree of mineralization and are ornamented with concentric striae.
- Extremely rare copepods.
- Above all by the ostracods, which are of great geological interest and are frequently studied because of their abundance and ubiquity.

1. LIVING OSTRACODS

General Organization

Members of the class Ostracoda have bodies that appear unsegmented. The head, thorax and appendages are ill defined and are contained in a calcitic **carapace**, the two **valves** of which are linked dorsally by an elastic **ligament** and a **hinge** (Figs 4.1 and 2). The body hangs inside the carapace like a sac. It is attached in the dorsal region and is fixed laterally to the valves by muscles. There are seven pairs of **appendages** – three pre-oral and four post-oral. They serve as sense organs and for the capture and mastication of food, for locomotion, and for cleaning the internal cavity.

Ostracods possess a digestive system, complex genital organs, a central nervous system, and there is frequently a median eye within the carapace behind a transparent furrow or **tubercule**.

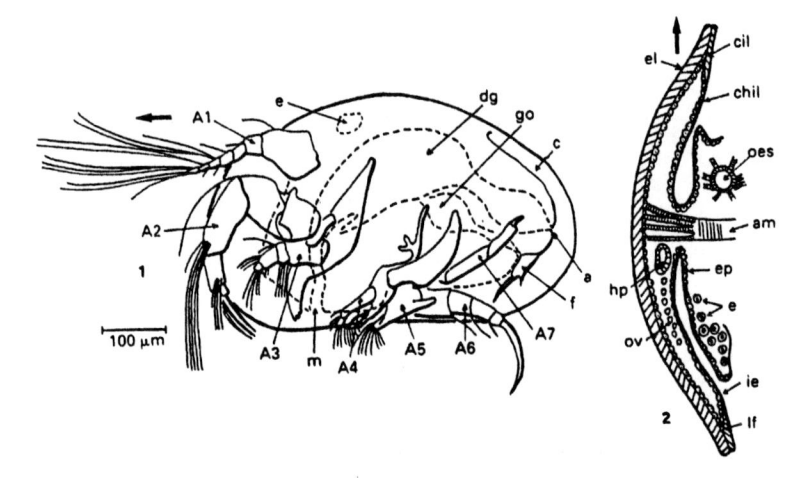

Fig. 4.1 Lateral view of a podocopid ostracod, without the left valve: A1–7 = appendages; a = anus; c = carapace; dg = digestive system; e = eye; f = furca; go = genital organs; m = mouth (×100). After Kesling (1951, figs 2 and 3)

Fig. 4.2 Median section of the left part of the body of an ostracod showing the structure of the two lamellae and the arrangement of the calcified parts (hatched) of the carapace: am = adductor muscle; cil = calcified internal lamella; chil = chitinous internal lamella; e = eggs; el = external lamella; ep = epidermis; hp = hepatopancreas; ie = internal edge; lf = line of fusion (concrescence); oes = oesophagus, ov = ovaries (×100). After van Morkhoven (1962, fig. 2)

Biology

The sexes are divided and dimorphism is common. The males are between three and ten times less numerous than the females and may be entirely absent when parthenogenesis is the rule. Reproduction takes place from both fertilized and unfertilized eggs. The eggs of lacustrine species are very resistant to desiccation while, on hatching, the nauplius larva is already equipped with a carapace and three pairs of appendages.

Growth is discontinuous and is marked by eight successive moultings. The larvae have anatomical characteristics that are less complex than those of the adult but in both cases the way of life is the same. Freshwater species grow to adulthood in about 30 days and live no more than a few months. Marine species reach maturity in periods varying from a few weeks to 3 years.

Ecology

Although some species are parasitic or commensal, most ostracods lead an independent existence. They inhabit every aquatic environment – oceans, coastal zones, estuaries, lagoons, fresh waters (lakes, rivers, ponds, springs), even moist forest soils. Many varieties are benthic: whether swimming, crawling or burrowing, they prefer stiller waters where the muds or fine sands of the bottom are rich in organic material. Some, such as the myodocopids, are planktonic.

Their diet is very variable and includes debris of microscopic and macroscopic algae, small living prey, dead animals and ooze.

They are sensitive to salinity, the nature of the substrate, and to temperature. Certain species are restricted to marine environments, others to freshwater (*Darwinulina* and most Cypridacea). In general, freshwater ostracods have a rather thin carapace with a surface that is smooth or slightly punctate and an adont hinge. Marine species, on the other hand, possess a more robust carapace that is often ornamented. Finally, there are brackish-water species that may reach population densities of tens of thousands of individuals per square metre on the bottom and can adapt to great variations in salinity. An example is *Cyprideis torosa* whose carapace becomes covered with nodosities as salinity decreases.

2. THE CARAPACE AND ITS FOSSILIZATION

Fossilization

When an ostracod dies, the body and appendages disappear leaving only the carapace to be fossilized with its valves either joined or separated. Planktonic species, which are only slightly mineralized, are rarely preserved. There have been a few reports of individuals being preserved with their appendages intact as a result of phosphate or silicate incrustation or the formation of pyritic internal moulds. These exceptional discoveries confirm that the most ancient types of ostracod had an anatomy comparable to those of present-day individuals.

Ostracod fossils abound in sediments of all kinds but are particularly numerous in clays and marls. The carapaces can be studied in detail only when they have been extracted from the sediment. With Palaeozoic species, however, it may be sufficient to observe the exterior of the valves at the surface of the stratification planes. If individuals are embedded in limestone, observation in thin section is not enough. In such cases it is sometimes possible to extract them by using acids to attack the cement.

Morphology of the Adult Carapace

The size of the carapace is in the order of millimetres, between 0.15 and 20 mm long for living species but up to 80 mm for the Palaeozoic leperditicopids. It consists of two laterally elongated **valves** articulated dorsally by a **hinge** (Figs. 4.3–6). The valves are hardly ever symmetrical in relation to the junction plane. One, the left in the case of the podocopids, is larger than the other, and the profiles are often different. Together with the absence of growth striae, this feature differentiates ostracod carapaces from Lamellibranch protoconchs.

The surface of the valves can be smooth, punctate or ornamented with a

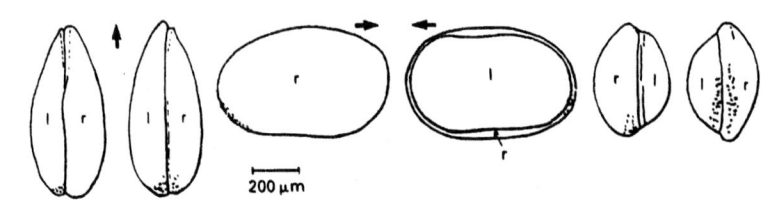

Fig. 4.3 Different views of the same carapace of *Cytherella*. From left to right: dorsal, ventral, right lateral, left lateral, anterior and posterior. l = left valve; r = right valve (×35). After Herring (1966, fig. 26)

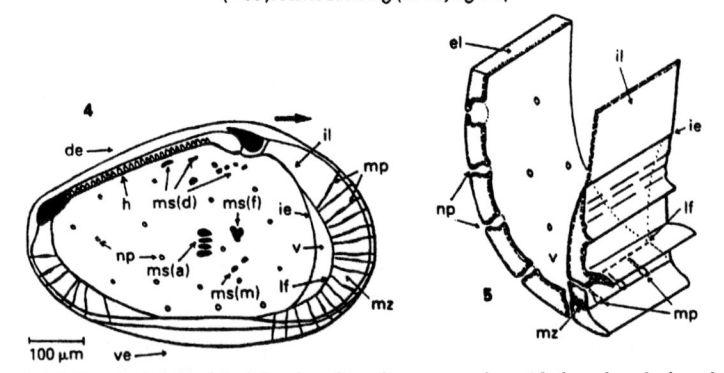

Fig. 4.4 Internal view of the left valve of a cytheracean podocopid: de = dorsal edge; el = external lamella; h = hinge; ie = internal edge; il = internal lamella; lf = line of fusion; mp = marginal pores; ms(a) muscle scar (adductor); ms(d) = muscle scars (dorsal); ms(f) muscle scars (frontal); ms(m) = muscle scars (mandibular); np = normal pores; v = vestibule; ve = ventral edge (×90). After van Morkhoven (1962, fig. 3)

Fig. 4.5 Details of the ventral edge of a valve; the chitinous non-calcified parts are shown by the dotted lines; abbreviations as in Fig. 4.4. After van Morkhoven (1962, fig. 32)

network of ribs, spines or aliform (wing-shaped) extensions. Lobes and grooves are also found, particularly in Palaeozoic carapaces (see Figs 4.7, 10 and 12) the ventral edge of which may be ornamented with a velum (Fig. 4.13).

In the living animal, each valve is formed with a tegumentary 'duplicature' of two **lamellae** (Fig. 4.5):

- one external and entirely calcified;
- the other internal and chitinous except (at least for most of the podocopids) towards the periphery where it is calcified.

Where there is one, the calcified **internal lamella** forms the marginal zone as a lining to the external lamella. Beyond the line of fusion (or **line of concrescence**) forming the boundary of the marginal zone, the calcified portion of the internal lamella may extend to form a **vestibule** situated between the line of fusion and the internal edge.

Running perpendicularly through the external lamella, there are normal (or lateral) **pore canals**, which open at the exterior through a pore that may take the form of a sieve. In the living animal, these are occupied by hair-like sensorial structures called setae. The marginal zone is traversed by **marginal pores** situated in the plane of the junction between the two lamellae.

In the centre of the valve, or slightly away from it towards the front, lies the **muscle scar field** (adductor, mandibular, etc). The number, form and arrangement of the scars all vary considerably.

Although the forms of hinge are very varied, they may be reduced to three basic types (Fig. 4.6):

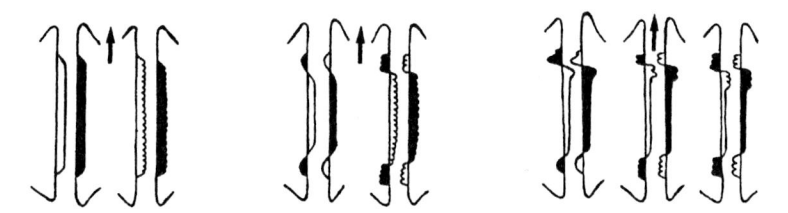

Fig. 4.6 Principal types of hinge: from left to right: two adont, two merodont and three amphidont types.

- **Adont**: the cardinal edge of one valve has a groove into which the bar of the other valve fits.
- **Merodont** or **taxodont**: the groove of one valve and the bar of the other are framed respectively by two teeth and two sockets.
- **Amphidont** or **heterodont**: the same as the merodont type with the addition of one tooth and one socket, which terminate respectively the anterior side of the median bar and groove.

All these cardinal elements – groove, bar, teeth and sockets – may be smooth, toothed or lobed.

Sexual dimorphism (see Figs 4.7, 10 and 25) is common. The carapace of the male is often more elongated but less high than that of the female. The female carapace has a posterior bulge and, in certain Palaeozoic genera (e.g. *Beyrichia* and *Craspedoboldina*) it contains an anteroventral brood pouch.

Larval Carapaces

The larval stages are distinguished from those of the adult by their smaller size, a rounded profile and the 'incomplete' appearance of the carapace, which is rudimentary in terms of valves, ornamentation, muscular scars, marginal zone and hinge. Sexual dimorphism does not appear until the very last stages. It has been possible to reconstruct the ontogenic development of several fossil species (Fig. 4.7).

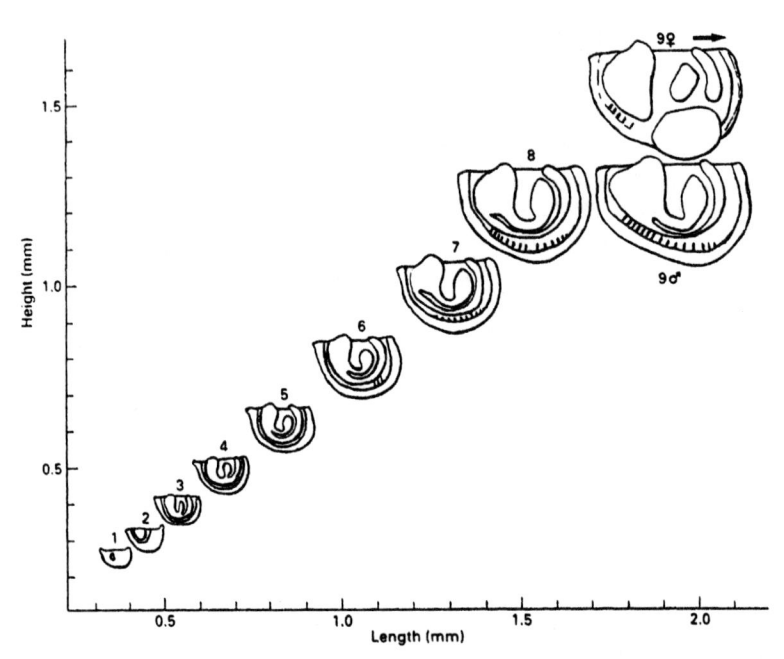

Fig. 4.7 Ontogeny of a palaeocopid ostracod, *Craspedobolbina clavata* (Kolmodin, 1869), from the Silurian of Gothland, based on a biometric examination of 3461 right valves; sexual differentiation appears only in the last stage (×13). After Martinsson (1962, figs 21 and 22)

Orientation of Carapaces

The hinge is carried by the dorsal edge. Looking at the lateral view from front to rear, the following characteristics can be seen:

- The anterior extremity is generally rounded whereas the posterior extremity is pointed.
- The muscle scars are situated in the anterior half of the valves and mandibular scars are in an anteroventral position.
- There are anterodorsal ocular markings.
- There are large spines and aliform extensions directed towards the rear.

Benthic varieties are seen in dorsal view to have an anterior part that is more compressed than the posterior. Finally, Palaeozoic ostracods have a rectilinear dorsal edge and a convex ventral edge while the lower part of the subventral groove S2 is directed towards the front.

3. SYSTEMATIC SURVEY

Although biologists are taking an increasing interest in the carapace, the classification of living ostracods is based on the morphology of the appendages. Every modern species has a characteristic carapace. If the external appearance of two different species is almost identical, they can readily be distinguished by study of the internal features.

As all the carapaces are constructed in the same way, specialists classify the 1000 genera and some 10 000 species of the group according to:

- The contour of the valves and the distinctive features of the ventral edge.
- The structure of the hinge.
- The characteristics of the marginal zone.
- The pores.
- The ornamentation.

In the class Ostracoda, the main divisions in the system of classification are interpreted somewhat differently by various specialists.

Order Leperditicopida

The leperditicopids (Fig. 4.9) are large in size and are characterized by a smooth surface and some 200 small circular muscle scars. They may form a distinct lineage within the ostracods.

Order Palaeocopida

The palaeocopids or beyrichicopids (Figs 4.10–13) are similar to the preceding group in that they have a thick adont carapace without a marginal zone and with a rectilinear dorsal edge and sharp cardinal angles. They differ from the leperditicopids, however, in terms of their smaller size, ornamentation of lobes, groove, velum and frequent sexual dimorphism.

Order Myodocopida

The planktonic myodocopids (Figs 4.14 and 15) form a heterogeneous group with a carapace that is thin, slightly mineralized, adont, and with a very narrow marginal zone. They are distinguished either by an anteroventral slit from which antennae emerge (*Cypridina*) or by a deep dorsal groove (*Entomozoe*).

Order Podocopida

Most of the post-Palaeozoic and Recent genera (Figs 4.18–22 and 26–36) are contained within the vast suborder *Podocopina*. They have an arcuate dorsal edge and a hinge that is sometimes adont but more often merodont or amphidont. The marginal zone is well developed (except in *Darwinulina*) and the number of muscle scars is small (less than ten). Four major superfamilies are distinguished: Bairdiacea, Cypridacea, Darwinulacea and Cytheracea.

Members of the suborder *Platycopina* (Figs 4.3, 16 and 17) have a robust shell that is smooth and adont. It has a narrow marginal zone and a reniform field with biserial scars.

PART 1: MICROFOSSIL GROUPS

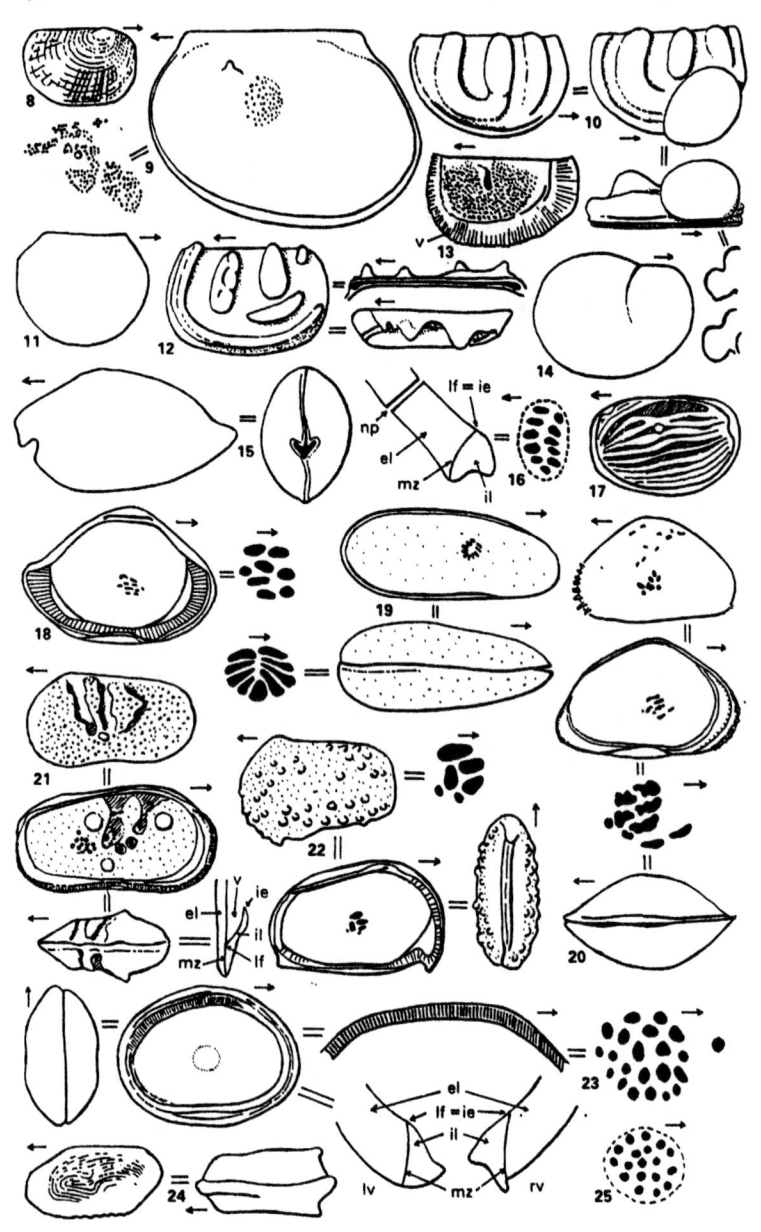

Finally, the suborder *Metacopina* (Figs. 4.23–25), a group intermediate between the two preceding ones, have, as their distinguishing feature, a circular field of some 20 muscle scars.

4. OSTRACODS THROUGHOUT GEOLOGICAL TIME

Emergence and Spread

Archaeocopida – generally recognized as primitive ostracods – are known to have existed at the beginning of the Cambrian. The bivalve carapace is flexible, slightly calcified, and has a rectilinear cardinal edge.

Fig. 4.8–25 Branchiopods (8); Leperditicopida (9); Palaeocopida (10–13); Myodocopida (14,15); *Podocopina* Platycopina (16,17); with adont hinge Bairdiacea (18), Darwinulinacea (19), Cypridacea (20–22) and Metacopina (23–25): ms = muscle scars; lv = left valve; rv = right valve

8 *Estheriella* (Triassic–Cretaceous) (×6)
9 *Leperditia* (Silurian–Devonian) (×2.5): left, ms
10 *Beyrichia* (Silurian): left, rv ♂; right, rv ♀ in lateral, ventral views and in serial sections; three successive lobes (from front to back: L1, L2 and L3) separated by two grooves (S1 and S2) (×10)
11 *Aparchites* (Ordovician–Devonian): rv in lateral section (×7)
12 *Drepanella* (Ordovician–Silurian): three lobes and two grooves; above and right, rv hinge (×12)
13 *Eurychilina* (Ordovician–Silurian): a dorsal groove; v = velum (×10)
14 *Entomozoe* (Silurian–Permian): a dorsal groove (×10)
15 *Cypridina* (Cretaceous–Recent): lateral and front views of the carapace, anteroventral incision (×15)
16 *Cytherella* (Triassic–Recent) cf. also Fig. 4.3. Left, section of the ventral edge of the lv: el = external lamella; ie = internal edge; lf = line of fusion; il = internal lamella; mz = marginal zone; np = normal pore. Right, ms (×130)
17 *Glyptopleura* (Carboniferous–Permian) (×18)
18 *Bairdia* (Ordovician? or Carboniferous–Recent): left, internal view of lv (×25); right, ms (×100)
19 *Darwinulina* (Carboniferous–Recent): carapace (×50); below and left, ms (×130)
20 *Cypris* (Eocene–Recent): from top to bottom, external and internal views of lv (×10); ms (×30); and ventral view of carapace (×10)
21 *Ilyocypris* (Jurassic–Recent): from top to bottom, internal and external views of lv (×30) and dorsal view of carapace (×25); below and right, section of the ventral edge of a lv (for legend see (9) above, plus v = vestibule)
22 *Cypridea* (Jurassic–Cretaceous): ventral view of carapace and lv, anteroventral rostrum (×30); above and right, ms (×60)
23 *Ogmoconcha* (Triassic–Lias): carapace (×25); lv (×35); hinge details (×50); ms (×150); below, section of ventral edge of lv and rv
24 *Quasillites* (Devonian–Carboniferous): lateral and ventral views of the carapace (×25)
25 *Euglyphella* (Devonian); ms (×35)

After Jones, Kesling, Scott, Sylvester-Bradley, Triebel, Ulrich, van Morkhoven and Wagner

PART 1: MICROFOSSIL GROUPS

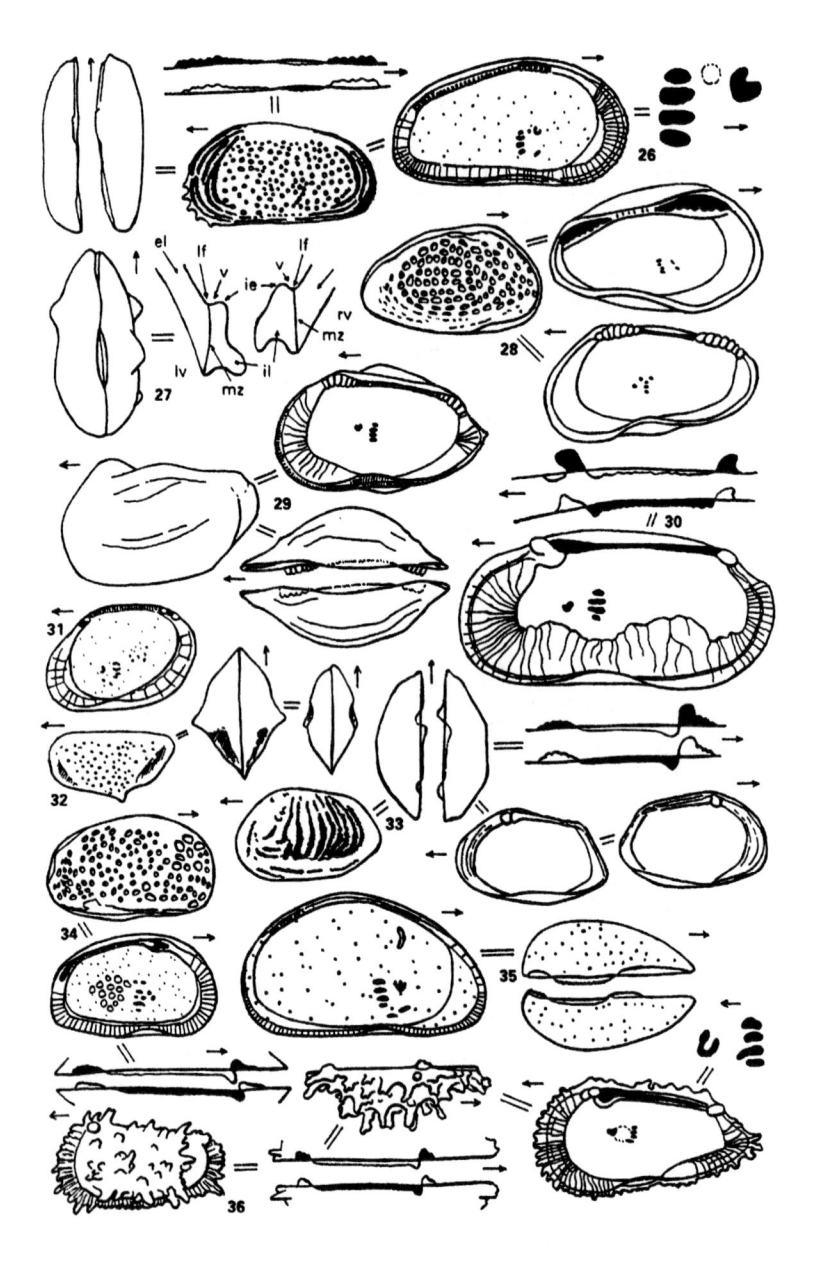

During the Ordovician, the large Leperditicopida and the first Palaeocopida flourished and then remained predominant until the Devonian. Pelagic limestones of the Devonian and the Lower Carboniferous contain rich associations of spiny shelled Myodocopida (Entomozoacea). Towards the end of the Palaeozoic, the Palaeocopida declined while the Platycopina, Metacopina, Bairdiacea and the first Cytheracea (*Monoceratina*) took on a certain importance.

The Triassic was a crucial period. Some groups of Palaeozoic ostracods such as the Platycopina and the Bairdiacea held their own but the Palaeocopida became extinct. Meanwhile new lineages of Cypridacea and Cytheracea emerged that would in future predominate.

The associations of the Liassic show little diversity (*Ogmoconcha*). Certain genera are confined to the marine facies of the Jurassic (*Procytheridea*) and the Lower Cretaceous (*Protocythere*). Lacustrine and brackish facies of the Purbeckian and Wealdian contain numerous Cypridacea (*Cypridea, Ilocypris*) associated with *Darwinulina* and with certain Cytheracea (*Macrodentina*).

Several Cretaceous genera continued into the Eocene. In the Cenozoic and Recent times, the Cypridacea are predominant in lacustrine environments and the Cytheracea in marine environments.

Throughout geological time, ostracods have shown the following evolutionary trends:

- Reduction in size.
- Modification of contour with the dorsal edge rectilinear and horizontal during the Palaeozoic but more or less convex and inclined subsequently.
- Growing complexity of the hinge, starting adont and then becoming merodont at the end of the Palaeozoic and amphidont from the late Jurassic.
- Reduction in the number of muscle scars from more than 20 in the Palaeozoic to less than 10 and even 4 subsequently.

Fig. 4.26–36 Cytheracean podocopid ostracods with a merodont or amphidont hinge and adductor muscle scars arranged in a vertical line
26 *Cytheridea* (Eocene–Recent): left to right, dorsal view of lv♂ and rv♀ (×30), external view (×30) and internal view (×50) of lv; ms (×30)
27 *Cyprideis* (Miocene–Recent): left, dorsal view of carapace with tubercles (×30); right, section of the ventral edge of the carapace (for legend see Fig. 4.)
28 *Procytheridea* (Jurassic) (×60)
29 *Protocythere* (Jurassic–Cretaceous); ♀ (×35)
30 *Cytheretta* (Palaeocene–Recent) (×50); above, hinge
31 *Loxoconcha* (Palaeocene–Recent) (×40)
32 *Monoceratina* (Devonian–Recent): from left to right, views of the carapace–lateral (×25), ventral (×25) and dorsal (×15)
33 *Macrodentina* (Jurassic–Cretaceous) (×30): above and right, hinge
34 *Hemicythere* (Oligocene–Recent) (×40): below, hinge
35 *Xestoleberis* (Cretaceous–Recent): left, internal view of lv (×70) and dorsal view of carapace (×50)
36 *Trachyleberis* (Eocene–Recent): left to right, lateral view of lv, dorsal and internal views (×25); below, hinge; right, ms

After Goerlich, Martin, Roth, Ruggieri, Sylvester-Bradley, van Morkhoven and Wagner

- Progressive calcification of the internal lamella, with progressive development of the marginal zone and vestibule.
- Appearance of normal cribrate pores from the Lias.
- Extension and increased complexity of the marginal pores, which acquire a flexuous branching appearance from the early Cretaceous onwards.

Palaeoecology

In the Palaeozoic, most ostracods were marine and benthic. Planktonic forms appear from the beginning of the Ordovician. The first lacustrine forms (*Darwinulina*) did not emerge until later in the Carboniferous.

The post-Palaeozoic forms show an ecological diversity comparable to those of today. Some are stenohaline, others are euryhaline. Ostracods multiplied in numbers sufficient to give rise to rocks consisting of little else ('*Cypris*' limestones of the Limagne Oligocene). Finally, abyssal species make it possible to follow the deep-ocean currents and the movement of the cold water circulations.

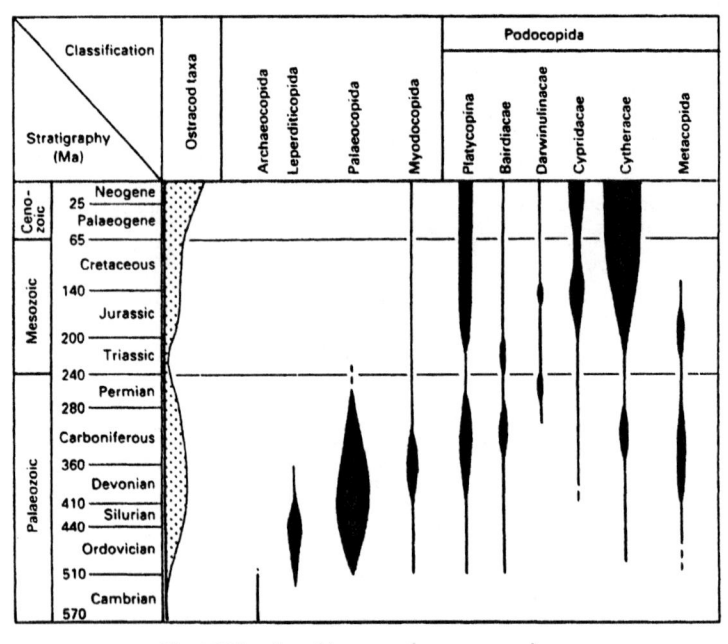

Fig. 4.37 Stratigraphic range of some ostracod groups

CONCLUSION

During the long period in which micropalaeontological examination was restricted to the residues left by washing, ostracods were the second group, in terms of geological interest, after foraminifera. Today, their importance in the field of stratigraphy is somewhat diminished, yet, because of their ecological diversity, they are tending to assume a dominant role palaeogeographic reconstructions, particularly of environments with variable salinity and deep water oceanic environments.

BIBLIOGRAPHY

Readers may obtain additional information from R.H. Benson, J.M. Berdam, A. Van den Bold et al., Ostracoda, in R.C. Moore (ed.), *Treatise on Invertebrate Paleontology. Part Q. Arthropoda. 3: Crustacea* (Geological Society of America and University of Kansas Press, Lawrence, 1961); F.P. Van Morkhoven, *Post-Palaeozoic Ostracoda*, 2 vols (Elsevier, Amsterdam, 1962–3); and N. Grekoff, *Aperçu sur les Ostracodes Fossiles*, 2nd edn (Technip, Paris, 1970). The work of R.H. Bate, E. Robinson and L.M. Sheppard, *Fossil and Recent Ostracods* (British Micropalaeontological Society Series, 494pp., 1982) is a good introduction to the problems associated with this group of microfossils.

There are descriptions of all genera and species in B.F. Ellis & A.R. Messina, *Catalogue of Ostracoda* (American Museum of Natural History, New York, 1952–); more than 15 000 cards have now appeared.

Some idea of the diversity of ostracod associations to be found in Western Europe can be obtained from R. Bate & E. Robinson (eds), *A Stratigraphical Index of British Ostracoda* (Special Issues of the *Geological Journal*, No. 8); Seel House Press, Liverpool.

Chapter 5

Calpionellids and Related Microfossils

This heterogeneous category is formed from calcareous microfossils, the dimensions of which are very small, sometimes even approaching those of the nanofossils. Their systematic position is uncertain, and, apart from their size, they have nothing in common but the simplicity of their morphology. In shape, they are either spherical or urnlike, and may be found with or without aperture.

Microfossils of this type are studied for the most part in thin sections. Examination of the microstructure of the wall and the statistical evaluation of the different sections make it possible – at least in certain cases – to distinguish 'genera' and to avoid confusions with other microfossils (radiolaria, isolated chambers of foraminifera, gyrogonites from charophyta, etc).

1. CALPIONELLIDS

Urn-shaped microfossils called calpionellids (Figs 5.1–7) occur in abundance in the fine marine and pelagic limestones of Mesogaean regions dating from the end of the Jurassic and from the early Cretaceous. The test, termed the lorica, is fibroradial, calcitic and no more than 45 to 150 μm in length. Fifteen genera (some 40 species) are known. Of these, six are particularly interesting both for their frequency and for their usefulness in stratigraphy. They are distinguished from each other by the appearance of the aperture and collar in axial section. The collar may, in exceptionally well-preserved examples, have a cylindrical extension.

Calpionellids appeared suddenly at the end of the Jurassic. They derive from an ancestral form (*Chitinoidella boneti* Doben, 1963) of the middle Tithonian, which had a wall that may have been 'chitinoid' but was more probably calcitic and microgranular. From the beginning they were very abundant and remained – until the end of the Valanginian – the predominant element in marine microplankton. No conclusive examples have been found for the

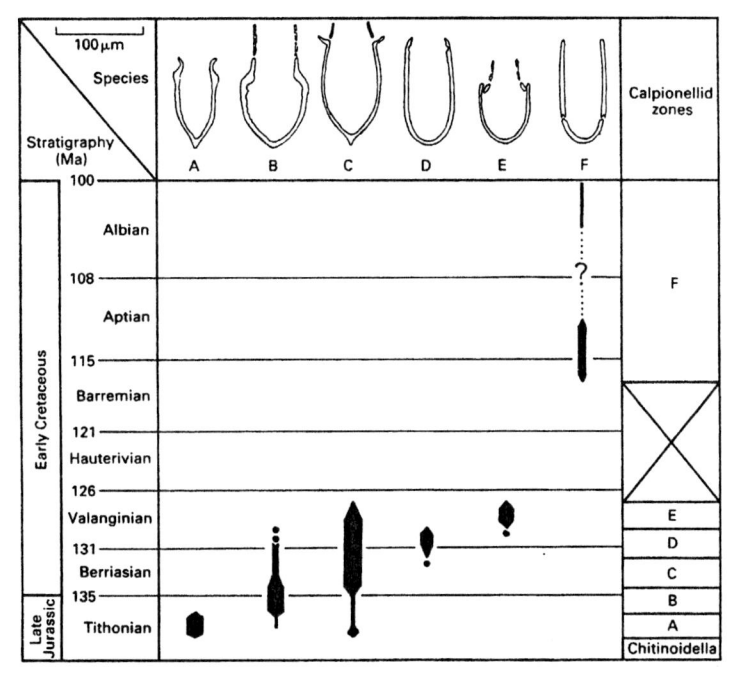

Fig. 5.1 Stratigraphic distribution of the following species of calpionellids: A, *Crassicollaria intermedia* (Durand-Delga, 1957); B, *Calpionella alpina* Lorenz, 1902; C, *Tintinnopsella carpathica* (Murgeanu & Filipescu, 1933); D, *Calpionellopsis* spp.; E, *Calpionellites darderi* Colom, 1934; F, *Colomiella* spp. The axial sections are magnified about ×150. After data from J. Remane.

Hauterivian. They reappear once more in the Barremian but die out completely at the end of the Albian.

The palaeogeographic distribution of calpionellids was wide. They are found in the Mediterranean, the Persian Gulf and in the Caribbean but are lacking in the Indo-Pacific domains. They were marine microorganisms and the evidence of other microfossils with which they are associated (coccoliths, *Nannoconus*, radiolarian and planktonic foraminifera) indicates that they were certainly pelagic.

The calpionellids have often been classified as ciliate protozoa (phylum *Ciliophora*) together with tintinnids. The latter are planktonic microorganisms found in present-day seas, estuaries and lakes, and of which fossil remains are occasionally found in lagoonal series dating from the Cretaceous and possibly from the Triassic. The reason for classifying the two groups together lies in partial similarities of size, morphology and ecology. There remains, however, a fundamental difference between the lorica of calpionellids and that of

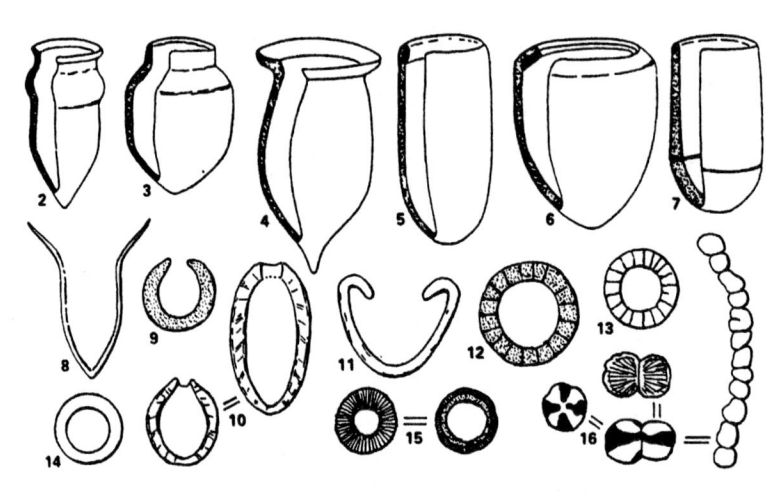

Fig. 5.2–16 Calpionellids (2–7), calpionellomorphs (8–11) and calcispheres (12–16)

2 *Crassicollaria intermedia* (Durand-Delga, 1957) (late Tithonian): oral ring at the base of the collar, maximum diameter at the ring (×280)
3 *Calpionella alpina* Lorenz, 1902 (late Tithonian–early Valanginian): straight collar, cylindrical and narrower than the lorica (×280)
4 *Tintinnopsella carpathica* (Murgeanu & Filipescu, 1933) (Late Tithonian–Valanginian): collar almost perpendicular to the lorica (×280)
5 *Calpionellopsis simplex* Colom, 1939 (Berriasian–early Valanginian); no separate collar, internal ring (×280)
6 *Calpionellites darderi* Colom, 1934 (Valanginian): bifidate collar directed diagonally towards the interior of the lorica (×280)
7 *Colomiella recta* (Bonet, 1956) (Aptian–Albian): straight cylindrical collar separated from the lorica by an articular surface (×280)
8 *Vautrinella* (Devonian) (×70)
9 *Cadosina* (Jurassic–Cretaceous): imperforate microgranular wall (×300)
10 *Pithonella* (Cretaceous): imperforate fibroradial wall, two different species (×200)
11 *Bonetocardiella* (Cretaceous): aperture at the bottom of a depression (×200)
12 *Pachysphaerina* (Carboniferous): perforate(?) microgranular wall (×70)
13 *Cytosphaera* (Carboniferous): perforate fibroradial wall (×50)
14 *Calcisphaerula* (Cretaceous): imperforate fibroradial wall (×150)
15 *Stomiosphaera* (Jurassic–Cretaceous): bilamellar wall, two different species (×250)
16 *Globochaete* (Silurian–Cretaceous): left to right, isolated individual seen in polarized light: two individuals seen in natural light (above) and polarized light (below); individuals in a chain (= *Eothrix*) (×130).

After Andri, Borza, Conil & Lys, Cuvillier, Lezaud, Remane, etc.

tintinnids: the former is rigid and calcitic whereas the latter is flexible, chitinous and often agglutinated. The currently prevailing view is that the two groups do not have a common line of descent. It is thought that calpionellids are a type of protozoa with unknown affinities and a restricted distribution in time.

2. CALPIONELLOMORPHS

Calpionellomorphs (Figs 5.8–11) are similar to calpionellids in terms of morphology, having a lorica that is spherical or ovoid in shape and always contains an aperture. They are distinguished from calpionellids, however, by the microstructure of the wall.

Vautrinella, found in the Devonian 'griotte' limestones of the Sahara and southern France, has a somewhat large lorica (200 to 250 μm) that exhibits a pronounced flaring of the collar. Other smaller individuals have been collected in the Silurian and the Ordovician of the Sahara. Found in association with cephalopods, the vautrinellids lived in the sea and were probably pelagic. Although they have been considered as genuine calpionellids, this identification must be regarded as premature given our lack of knowledge with regard to the microstructure of the lorica, which is invariably recrystallized, and the significant stratigraphic hiatus separating them from the true calpionellids.

During the Mesozoic, calpionelliomorphs are represented by specimens that are small in size, ranging between 50 and 120 μm in length. Three genera in particular (*Cadosina*, *Pithonella* and *Bonetocardiella*) are worth citing in view of their abundance in a variety of fine or chalky limestones. Both their wide geographical distribution and their frequent association with radiolaria and calpionellids suggest that they lived as plankton in the high seas or on the continental shelf.

3. CALCISPHERES

Calcispheres (Figs 5.12–16) are hollow spheres with a diameter that is generally less than 100 μm. The calcareous wall is somewhat thick and lacks an aperture.

Calcispheres are extremely common in the neritic limestones of the late Devonian and the early Carboniferous where they are found in association with benthic foraminifera, calcareous algae, brachiopods and echinoderms. The 'genera' are distinguished from each other by the presence or absence of perforations and the thickness and microstructure of the wall; these characteristics are, however, dependent on fossilization. Palaeozoic calcispheres have sometimes been regarded as primitive fusiline foraminifera (Parathuramminidae) or as the cysts of marine green algae (Dasycladales).

The Tithonian limestones of the Jurassic and the early Cretaceous, which contain radiolaria, calpionellids and planktonic foraminifera are also rich in *Stomiosphaera*. This probably planktonic calcisphere has a wall with two layers, the inner one microgranular and the outer fibroradial. In the late Cretaceous, calcispheres are represented by *Calcisphaerula*.

One last form, *Globochaete*, is commonly found in association with various planktonic microfossils in Palaeozoic (from the Silurian) and Mesozoic limestones. These minute (from 70 to 200 μm) fibroradial calcitic spheres occur both in isolation and arranged in rows. It is assumed that they are shells that filled with calcite in the course of diagenesis.

CONCLUSION

These remains illustrate well the paradoxical position in which micropalaeontologists are often placed. They are dealing with microfossils whose systematic classification is uncertain but which are, nevertheless, of considerable geological interest as they play a notable part in lithogenesis and provide valuable stratigraphic and palaeogeographic information.

BIBLIOGRAPHY

Supplementary information can be found in J. Remane, *Annales Guébhard-Séverine*, Neufchâtel 47, pp. 369–393 (1971); K. Borza, *Die Mikrofazies und Mikrofossilien des Oberjuras und der Unterkreide der Klippenzone der Westkarpaten* (Slovak Academy of Sciences, Bratislava, 1969); and R. Conil & M. Lys, *Mémoires de l'Institut Géologique de Université de Louvain* 23, 28–52 (1964).

Chapter 6

Mineralized Plant and Animal Remains

It has already been stressed that the frontiers of micropalaeontology are not sharply defined and that they are, to a large extent, shaped by tradition. Micropalaeontologists may well be involved in subsidiary studies of certain macrofossils (e.g. sponges, bryozoans, tubes of serpulid worms), dwarf macroorganisms (e.g. crinoids, gastropods) or juvenile stages of macroorganisms (e.g. mollusc protoconchs, echinoderm pluteus, trilobite protapsis). However, more important to micropalaeontologists are the remains of calcareous algae, pteropod molluscs, tentacularids and small skeletons.

1. CALCAREOUS ALGAE

This term is used to designate multicellular photosynthetic organisms that fossilize because they possess a calcified **thallus**. The external form of the thallus may be distinctive but recognition usually proceeds from an examination of the calcareous material, which, during the life-cycle of the organism, encrusts the walls of coenocytic filaments or cells.

Calcareous algae appear in four main forms:

(1) Segments of **Dasycladales** (Chlorophyta: class Chlorophyceae) (Figs 6.1–8): initially of aragonite, they are usually recrystallized as calcite. During the life of the organism, they are external, moulding all or part of the principal axis, the branches and the gametangia, which do not fossilize. Known from the Cambrian onwards, these calcareous green algae have always lived in shallow (usually less than 5 m) tropical and warm-temperate marine waters which are close to the shore and sometimes of abnormal salinity.

(2) Assemblages of calcareous tubes with a diameter of less than 100 μm and usually without septae. They are characteristic of the blue-green algae or cyanobacteria (**Cyanophyta**) (Figs 6.9 and 10) and the **Udoteaceae** (Chlorophyta) (Figs 6.11–13). The cyanophyte remains take the form of non-

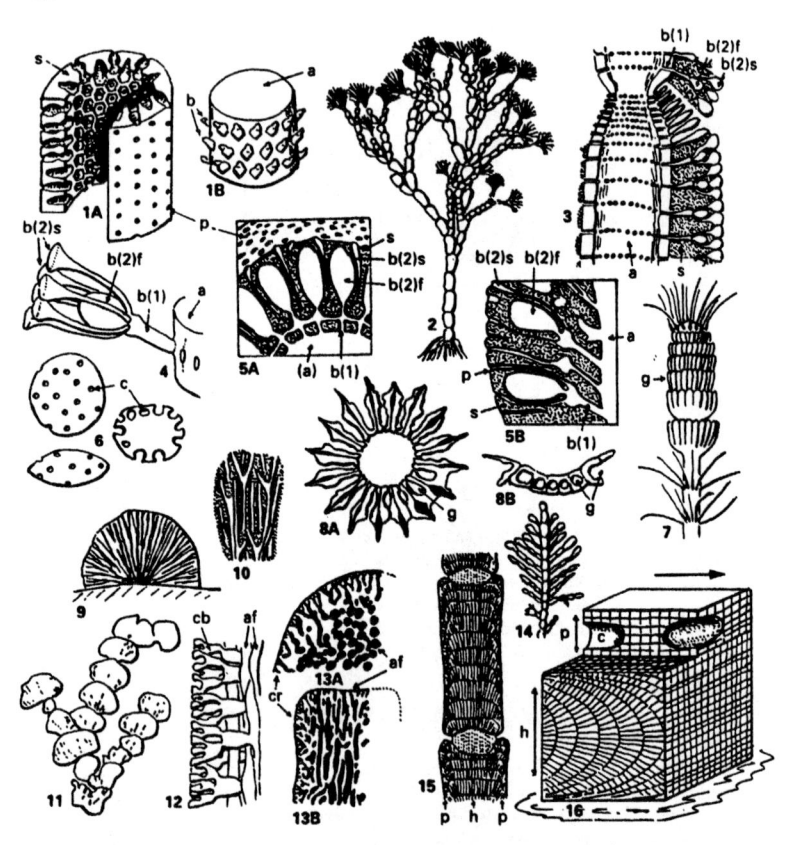

Fig. 6.1–16 Algae with a calcareous segment surrounding the axis: dasycladacean chlorophytes (1–8); algae with a calcareous sheath around the filaments: cyanophytes (9–10) and udoteacean chlorophytes (11–13); algae with calcified cellular walls: corallinacean rhodophytes (14–16)

1 Diagram of a non-articulated dasycladacean (*Koninckpora*, Carboniferous): a calcareous segment (A) surrounds the axis (a) and the stumps of the branching verticilli (b), which are without visible gametangia (endospore mode) (×5)

2–5 Dasycladales with articulated thallus (*Cymopolia*, Cretaceous–Recent): 2, external view of the thallus of a living individual (×0.5); 3, axial section of two segments (×8); 4, gametangium at the end of the primary branch (cladospore mode), surrounded by six secondary sterile branches, without the calcareous segment (×80); 5, calcareous segment in transverse section (A) and axial section (B) (×45): a = axis (or its emplacement); s = calcareous segment (shaded); p = pores; b(1) = primary branch; b(2) secondary branch; b(2)f = fertile secondary branch; b(2)s = sterile secondary branch

6 Isolated gametangium of the dasycladacean *Terquemella* (Eocene): calcareous filling

branching filaments or trichomes (*Girvanella*), which may occur in isolation or arranged in loose bundles within stromatolites of which mention will be made later. Alternatively, the filaments may be branching and arranged in the form of encrusting pads. The udoteaceans have an upright thallus that branches and may or may not be articulated. Dating from the Cambrian, or at least the Ordovician, they are algae of shallow (a few metres deep) tropical waters of normal salinity. As a rule, the closer the filaments are to the periphery of the thallus, the more calcified their wall (magnesian calcites for the cyanophytes, aragonite for the udoteaceans).

(3) The thalluses of red algae or **Rhodophyta** (Figs 6.14–16) especially cryptomerial: formed from the juxtaposition of rows of polyhedral cells in the order of tens of micrometres, the walls of which are impregnated with magnesian calcite. They may be crustose (**Solenoporaceae: Lithothamniae**) or erect and articulated (**Corallinaceae**). The rows of cells at the centre are of unlimited growth and they form the **hypothallus**. Secondary peripheral rows of limited growth extend laterally to form the **perithallus**. All rhodophyte fossils are marine. The Solenoporaceae are known from the Cambrian to the Eocene; the Corallinaceae, represented by primitive forms in the Palaeozoic, appeared definitively only towards the end of the Jurassic but saw a significant development from the middle Cretaceous.

(4) Remains of Charophyta (Figs 6.17–27). Known since the Devonian and possibly the Silurian, they form a class that is well defined by test morphology and by a habitat of calm, shallow fresh water. Certain living species have adapted to hyposaline environments (estuaries and lagoons) while others are found in hyperhaline environments (saltmarshes). Some fossil species shared

---

except at the site of cysts (c), which produce the gametes; the thallus is unknown but probably non-calcified segment; on the left, external views; on the right, axial section (×60)

7,8 Another articulated dasycladacean (*Clypeina*, Jurassic–Eocene): the calcareous segment is reduced except around the verticilli of the gametangia (g). 7, Reconstruction of the thallus of an Eocene species (×8); 8, transverse section (A) and axial section (B) of a verticillus of gametangia belonging to a Jurassic species (×15)

9,10 Cyanophyte with nodular thallus and pluricellular filaments (*Ortonella*, Silurian–Carboniferous, or *Cayeuxia*, Jurassic–Cretaceous): 9, axial section of thallus (×10); 10, detail of branching filaments surrounded by limestone (×50)

11–13 Udoteaceans with articulated thallus and coenocytic filaments = siphons (*Halimeda*, Cretaceous–Recent). 11, External view of the thallus of a living individual (×1); 12, Detail of part of a decalcified plate; right, some axial filaments (af); left, some cortical branches (cb) (the intersiphonary space was filled with limestone) (×100). 13, Transverse section (A) and axial section (B) of a fossilized segment; the calcareous tissue is transparent and the site of the siphons is filled with sediment (×15)

14,15 Corallinaceae with erect and articulated thallus (*Corallina*, Eocene–Recent); 14, fragment of the thallus of a living individual (×1); 15, axial section of several segments separated by non-calcified sutures, hypothallium (h) with lines parallel to the axis of growth and perithallium (p) with perpendicular lines (×20)

16 Corallinaceae with crustose thallus (*Lithophyllum*, Cretaceous–Recent): portion of the thallus with hypothallium (h) and perithallium (p) pitted with conceptacles (c); the arrow indicates the axis of growth (×150)

After Genot, Johnson, Morellet, Steinmann, Wilbur, etc. and original drawings.

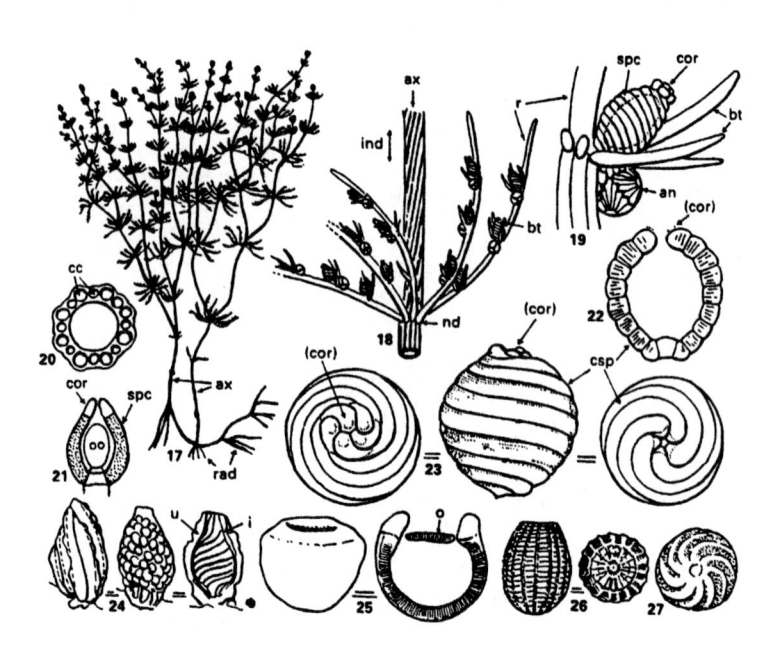

Figs 6.17–27 Charophyte calcareous algae: an = antheridia; ax = axis = cladome; b = branch; bt = branchlet = bracteal; cc = cortical cell; cor = coronule; i = involucre; ind = internode; nd = node; o = operculum; oo = oocyst; rad = radicle; spc = spiral cell

17–21 A living individual (*Chara*, Eocene–Recent): 17, external view of the thallus (×0.25); 18, detail of a verticillus of a branch carrying branchlet verticilli, antheridia and oogonia (×7); 19, details of a branchlet verticillus (×25); 20, transverse section of an axis (×25); 21, diagram of an oogone with an involucre formed from a verticillus with spiral cells (= specialized branchlets) surmounted by the coronule and enveloping the oocyst

22 Axial section of a gyrogonite with an apical pore (*Porochara*, Jurassic–Palaeocene) (×20)

23 Gyrogonite of *Tectochara* (late Cretaceous–Recent) with five sinistral spiral cells that join at the summit: left to right, views from above, profile and below (×20)

24 Gyrogonite of *Clavator* (late Jurassic–early Cretaceous): left to right, views of profile with utricle (u) utricle removed to make visible the nodular spiral cells and section (×25)

25 Umbellinid (Devonian–Carboniferous): left to right, profile view (×70) and axial section (×45) with wall without radial division and aperture closed by operculum

26 Gyrogonite of *Sycidium* (Silurian ? or Devonian–Carboniferous) with 18 vertical rows of cells: seen from left to right in profile and from above (×30)

27 Gyrogonite of *Moellerina* = *Trochiliscus* (Devonian) with nine cells forming a dextral spiral, seen from above (×20)

After Bourrelly, Bykova, Grambast, Harris, Migula, Peck, Reitlinger, Sachs, etc.

such habitats but female **gametangia** occur in greatest abundance in lacustrine sediments (e.g. clays, limestones and siliceous limestones) of the Mesozoic and the Cenozoic. These female organs, which are also called oogonia or nucules, are formed in the living plant from an oocyst, an involucre or cortex, and a coronule. The only part to fossilize is the calcite-impregnated internal wall of the five anticlockwise spiral cells of the involucre: this is the **gyrogonite**, which is usually bare but which, in the Clavatoraceae of the Jurassic and Cretaceous, is lined with a supplementary layer or **utriculus**. In the Devonian and the Carboniferous, charophytes produced gyrogonites with more varied forms. Apart from umbellinids of uncertain classification, mention must be made of the gyrogonites of *Sycidium*: these are formed from 16 to 20 rows of cells arranged vertically, or from cells following numerous left and right-handed spirals.

2. PTEROPODS AND TENTACULITIDS

Pteropods

These marine gastropods (Figs 6.28–31) in the subclass Opisthobranchia are adapted to a pelagic life. The foot is divided into two fins or parapodia which beat like wings in front of the body. Some varieties (in the order Thecosomata) have an aragonitic shell with a length ranging usually from 2.5 to 10 mm but which may reach 30 mm. The shell is either a sinistral trochospiral or a rectilinear or arcuate horn with bilateral symmetry.

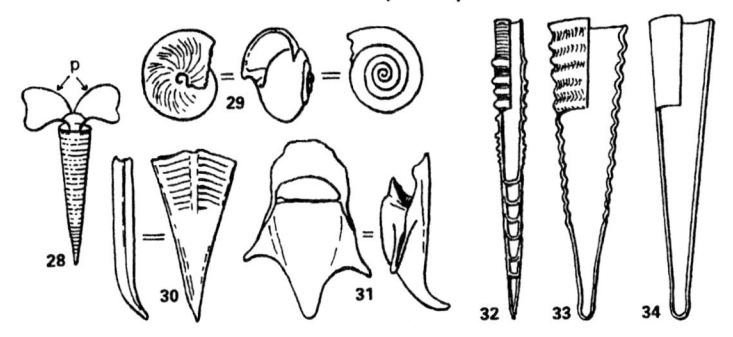

Figs 6.28–34 Pteropods (28–31) and tentaculitids (32–34)

28 *Hyalocylis*: living individual; p = parapodia (×3)
29 *Limacina* = *Spiratella* (Eocene–Recent) (×13)
30 *Clio* (Miocene–Recent) (×3)
31 *Cavolinia* (Miocene–Recent) (×4)
32 *Tentaculites* (Silurian–Devonian) (×3)
33 *Nowakia* (Devonian) (×13)
34 *Styliolina* (Devonian) (×13)

After Bé, Gilmer, Lyashenko, etc.

The 30 species that exist today are ubiquitous and live in tropical waters. Some are adapted to cold waters. Although they occur at all depths, they are mostly found between 300 and 500 m. The accumulation of their shells on the floor of the Mediterranean, the Red Sea and a few parts of the Indian and Atlantic oceans has given rise to pteropod oozes.

Uncommon in the fossil record, pteropods generally take the form of an internal mould. None has been identified with certainty before the Eocene but they are sometimes found in abundance in Neogene and Quaternary sediments.

Tentaculitids (or Cricoconarids)

Tentaculitids (class Tentaculitoidea or Cricoconarida) take the form of narrow cones with a tip that may be sharply pointed or bulbous (Figs 6.32–34). In size, they range from 0.8 to 80 mm with an average of 20 mm. The wall is calcitic and lamellar, and the interior is sometimes divided by floors. The external surface may be smooth, or with transverse rings or corrugation, and occasionally with longitudinal striation. They are known in the Silurian and especially the Devonian. Some 30 genera and several hundreds of species have been described. Distribution is world-wide. They are found in abundance in argillaceous and calcareous marine facies but are less common in sandstones and dolomites. The structure of the tentaculitids, their deposition patterns and associated fossils, all combine to suggest that some of the larger forms with thick walls and floors (*Tentaculites*) may have been benthic whereas the smaller thin-walled types (*Nowakia* and *Styliolina*) may, like present pteropods, have had a planktonic existence.

The organisms that produced these shells remain unknown although occasional internal moulds give indications that there would have been bilateral symmetry. Therefore, the systematic position of tentaculitids is uncertain. Some authors regard them as primitive pteropods, glossing over the immense gap that exists between the end of the Devonian and the appearance of the first true pteropods. The only basis for such a hypothesis is morphological convergence. Although they have been grouped here arbitrarily alongside the pteropods, there is every likelihood that the tentaculitids represent a different and original group, probably within the phylum Mollusca.

Figs 6.35–63 Isolated organic elements of macroorganisms
35–44 Spicules of siliceous sponges (hexactinellids and demosponges) (35–39 and 41–44) and calcareous spicule of calcisponge (40): 35, monaxon spicules; right, monactine, left, diactine (×30); 36, transverse sections of spicules; left, reniform; right, circular (×30); 37, globular spicules (*Rhaxella*) (×50); 38, triaxon spicule, triactine (×30); 39, triaxon spicule, hexactine (×30); 40, tetraxon spicule (×30); 41, two examples of microscleres (×150); 42, two irregular spicules (= desmas) of lithistid demosponge (×12); 43, five examples of microscleres (×150), 44, desmas arrangement typical of lithistids (×15)
45 Calcareous spicule of *Chancelloria* (Cambrian); the large axial cavity is compartmented and opens on to the exterior via a pore (×15)
46,47 Aragonitic spicules of ascidian; 46, polycitorid disc seen in profile (left) and from above (right) (×50); 47, *Micrascidites* of Didemnidae (×300)

MINERALIZED PLANT AND ANIMAL REMAINS

48 Aragonitic spicule (*Micralcyonites*) of alcyonarian coral (×70)
49 Calcitic spicule of a brachiopod (×75)
50–55 Calcitic sclerites of holothuroid echinoderms displaying various forms: circle (50), disc (51), grill (52), turret (53), anchor (54,55) (×1000 on average)
56 Brachial piece ('vertebra') of ophiuroid echinoderm seen from different angles (×15)
57 Planktonic crinoid echinoderm (*Saccocoma*): the arms are very long and the blades at the base are used for swimming (×5); known as *Lombardia* in wash residues (above and left) and thin sections (seriate sections below and left) (×25)
58 Two plates from the wall of a cirriped crustacean (*Balanus*, Eocene–Recent) (×1)
59 Transverse section of a plate of *Balanus* with channels separated by 'epithelial lamina' opening on to the exterior via slits (s) (×6)
60 Rhyncolite (calified cephalopod beak) formed from a stem and cap (×20)
61 Aragonitic fish otolith (×50)
62 Phosphatic fish tooth (×10)
63 Tooth of hamster (*Cricetus*), worn-down table with projecting enamel folds (%×5)

After Chaline, Deflandre, Frizzell & Exline, Hess, Jaeckel, Monniot, Moret, Rioult, Sdzuy, Sigal, Trejo, Verniory, etc.

3. ISOLATED ORGANIC ELEMENTS

Skeletal elements of small size and definite form are often found in thin sections and wash residues (Figs 6.35–63). Many multicellular animals have a skeleton formed from small pieces that are isolated in the living creature or easily dissociated on fossilization. These sometimes form a considerable part of fossil assemblages (taphocoenoses). Examples are spicules and sclerites of various forms: siliceous elements of the **hexactinellid** sponges (microscleres and particularly macroscleres); aragonitic elements of **ascidean** urochordates (known from the Dogger); and calcitic elements of the calcisponges, the *Chancelloria* of the Cambrian, the alcyonarian corals (from the late Cretaceous), some brachiopods and the **holothuroid** echinoderms (from the Carboniferous and possibly the Devonian). Within each group, these elements have a characteristic morphology although this is insufficient for correct determination of a species or genus. The same applies to plates of cirriped crustaceans, **echinoderm** ossicles (often represented by crinoid entrochs), the brachial extensions used for swimming by the crinoid *Saccocoma* (known in the late Jurassic and the early Cretaceous as *Lombardia*), brachial fragments ('vertebrae') of ophiuroids, pedicellariae and small radial plates of echinoids, etc.

For the sake of completeness, the following should be added:

- Rhyncolites or calcified beaks. These range in length from 2 to 50 mm and it is thought that they derive, for the most part, from shell-less cephalopods.
- Various remains of fish: teeth and phosphatized dermal scales, aragonitic otoliths of the internal ear (known from the Permian) ranging from 0.1 to 3 mm in diameter.
- The teeth of small mammals such as insectivores (shrews) and more particularly rodents (e.g. voles, squirrels, hamsters, dormice). Only one tooth per ton of sediment is found in palaeosoils and periglacial and fluviolacustrine deposits but they are more abundant in the debris of cave-floors and overhanging rock shelters. They are extremely useful for dating purposes and for reconstructing the continental palaeoenvironments of the Neogene and the Quaternary.

4. SKELETAL FRAGMENTS

Even where the standard fossils lack a definite form, it is still possible to study fragments of them in terms of general shape and microstructure.

Morphological Criteria

Where the fragments are relatively large, there is no great difficulty in attributing them to one or other metazoan group. Not only is it fairly easy to reconstruct the general form but there are often additional morphological details: perforations running through the shells of numerous brachiopods, channels dug in the shells of caprinid rudists (heterodont bivalves), 'epithelial lamina' from plates of the cirriped *Balanus*, etc.

In thin section, some of the fragments – even if they are quite small – have a normal appearance (*Cladocoropsis*, stromatoporoids, zoarium of bryozoans, serpulid tubes, shells of radiolitid rudists). Others, however, have quite unexpected forms such as *Lombardia* (Fig. 6.57) and the fine 'filaments' seen in Fig. 6.64. Rectilinear or arcuate, these are the sections of thin shells of pelagic bivalves found in fine limstones of the Devonian, Triassic and Jurassic.

Microstructural Criteria

Skeletal remains rapidly become fragmented and unrecognizable under the action of successive phases of friction. However, it is sometimes possible to identify them through their microstructure (Figs 6.65–67). Calcareous and phosphatic skeletal tissues have their own characteristic histology in the living creature, as has been shown for foraminifera; this also holds good for other groups of plants and animals.

Within a given group, the variations in microstructure may be very slight. Each fragment of an echinoderm skeleton is a single crystal of magnesian calcite formed into a trabeculate lattice. The skeletal tissue of madreporarian corals may be formed from acicular crystals of aragonite (width 0.3 to 2 μm and length up to 600 μm), which are arranged in bundles or sclerodermites. The calcified parts of udoteacean and dasycladacean chlorophytes also take the form of acicular crystals. The cell wall of corallinacean rhodophytes is formed from perpendicular rows of small globular crystals (width = 0.15 to 1 μm).

In molluscs, a dozen different types of microstructure have been discovered. The most common are:

- **Lamellar** types: superpositions of lamellae formed either from a single layer of flat crystals of aragonite (nacreous structure) or of calcite (foliate structure).
- **Prismatic** types: tight aggregates of prismatic columns formed from plane or convex layers of globular crystals.
- **Cross-lamellar** types: concentric structures of crystalline lamellae alternatively inclined in different directions.

Multiple variations of these types provide a sufficiently diversified histology of the shell for the identification of every mollusc family.

In principle, therefore, the examination of the microstructure of a tiny calcareous or phosphatic fragment makes it possible to identify its biotic origin and to discover the organism from which it derived. The minimal dimensions for the recognition of a fragment vary according to the method of observation and depend on the size of the basic microstructural unit at the scale in question: 100 to 75 μm for the light microscope, 15 and even 4 μm with the SEM.

The importance of these observations is unfortunately limited by the diagenetic modifications that accompany fossilization and which may mask or destroy the initial microstructural characteristics. Calcitic skeletons that serve to support secondary crystalline growths vary in the degree to which they are modified but usually remain recognizable (echinoderms). When aragonite is transformed into calcite, however, there is always a negative effect on

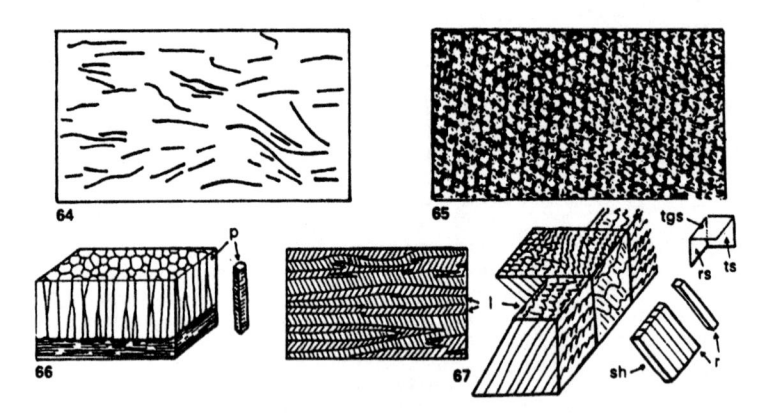

Figs 6.64–67 Skeletal fragments of macroorganisms
64 'Filaments' in a thin section of limestone (×10)
65 Fragment of the skeleton of a fossil echinoderm seen in thin section under a light microscope: the light areas correspond to the spongelike skeleton of biotic calcite; the dark areas correspond to the calcite mixed with impurities which, on fossilization, filled the cavities of the skeleton (×80)
66 Fragment of mollusc shell with a prismatic external layer and an internal layer that is either lamellar or nacreous: on the right, isolated prism (p) with growth striation (×100 on average)
67 Fragment of a mollusc shell, a microstructure displaying cross-bedding: left, tangential view under light microscope (×150); right, schematic diagram showing three contiguous sheets (sh) consisting of lamella (l) inclined to form an angle between them of 80°. Each lamella can be broken down into rods (r) with a diameter of 0.2 μm; radial section (rs), transverse section (ts) and tangential section (tgs)

After Cayeux, Cuvillier and original drawings

recognition: in a somewhat recrystallized limestone, fragments of udoteaceae, dasycladales, madreporarians and molluscs are virtually indistinguishable.

CONCLUSION

It can be seen, therefore, that the morphological and microstructural characteristics of certain macrofossils can be identified from wash residues, microfacies and nanofacies. The determination is, of course, far from precise and generally goes no further than the class or even phylum. Despite the lack of precision, these microfossils must be checked with care. They indicate the presence of organisms (e.g. holothuroids, ascidiaceans), which are found in sediments only in microscopic form. They establish their participation in lithogenesis and can provide valuable stratigraphic and palaeogeographic information.

BIBLIOGRAPHY

In this field, it is worthwhile referring to textbooks and manuals of classical palaeontology. J.L. Wray, *Calcareous Algae* (Elsevier, Amsterdam, 1977) gives an overview of fossil calcareous algae. Between 1958 and 1966 J.H. Johnson published in the *Quarterly* and the *Professional Papers of the Colorado School of Mines*, several clear and well-illustrated syntheses on Palaeozoic and Mesozoic algae, and on the rhodophytes. For the latter, see also the works of M. Lemoine and A.F. Poignant. The Dasycladales have recently been the subject of two memoirs: J.P. Bassoulet *et al.*, 'Les Algues Dasycladales du Jurassique et du Cretacé' (*Geobios Special Memoir* no. 2, 1978) and P. Génot, 'Les Dasycladacées du Paléocène supérieur et de l'Eocène du Bassin de Paris' (*Mémoires de la Societé géologique de France*, n.s., 138, 1980). The publications of L. Grambast and his students are a useful source of information on charophytes.

The standard works for studying fragments of macrofossils are O.P. Majewske, *Recognition of Invertebrate Fossil Fragments in Rocks and Thin Sections* (Brill, Leyden, 1969) and A.S. Horowitz & P.E. Potter, *Introductory Petrography of Fossils* (Springer, Berlin, 1971).

Monographs have been published on some of the groups cited: D.L. Frizzell & H. Exline 'Monograph of Fossil Holothurian Sclerites' (*Bulletin of the University of Missouri*, no. 89, 1955) and H. Lardeux (1969), 'Les Tentaculites d'Europe Occidentale et d'Afrique du Nord'. (*Cahiers Paléontologie*, Paris 1969).

Some idea of the microstructure of mineralized tissues of macroorganisms can be obtained by reference to W.W. Hay, S.W. Wise and R.D. Stieglitz, SEM study of fine grain size biogenic carbonate particles. (*Transactions of the Gulf Coast Association Geological Society*, 20, 287–302, 1970).

Chapter 7

Calcareous Nanofossils

Characterized by their minute size (⩾50 μm), nanofossils[1] are non-mineralized (e.g. certain pollens), siliceous (e.g. Chrysomonadales) or calcareous. The last group is the most important, being well represented in sediments and very widely used in stratigraphy. However, although it has been possible to establish their history and stratigraphic distribution, their systematic classification remains confused. Many forms belong to the Coccolithophyceae but the affinities of others are uncertain or unknown.

1. LIVING COCCOLITHOPHORES

General Organization

Coccolithophores (class Coccolithophyceae) are chrysophyte unicellular algae. The rigid cell ranges in size from 10 to 50 μm and may be globular, pyriform or fusiform (Fig. 7.1). It has a nucleus, two greenish-yellow or light-brown chloroplasts and, usually, a short whip-like appendage or haptonema. This is flanked by two flagellae of the same length, which emerge from an orifice ('mouth') situated at the anterior pole of the cell.

The pectic (mucilaginous) envelope of the cell is hard on the outside. Usually, there is a shell formed from tiny calcitic platelets or coccoliths embedded in this layer if it is thick, or resting on the surface if it is thin. The coccoliths number between 10 and 30 and may be either slightly separated or contiguous. In the latter case, they are either joined or partially overlapping. The shells that do not dissociate give rise to the coccospheres. Certain present-day species have coccoliths with different forms. The 'buccal' forms are more ornamented than the others.

[1] See footnote p.1.

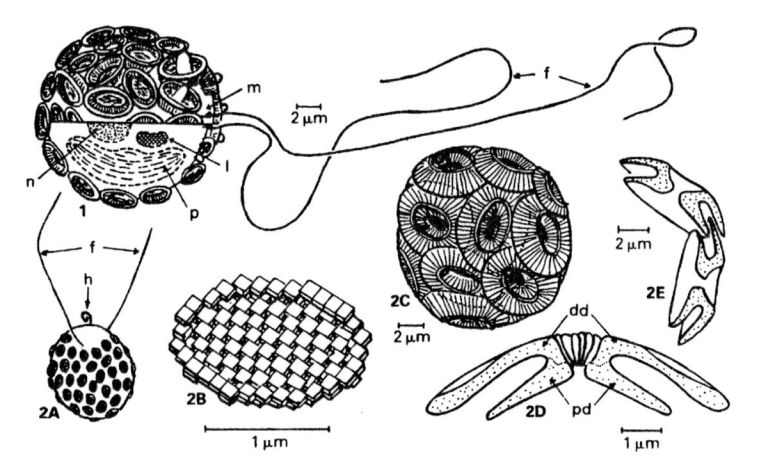

Fig. 7.1 Living coccolithophore (*Syracosphaera*). Flagellate stage with coccosphere formed from two types of coccoliths: normal discoliths with a flat base and 'buccal' discoliths with a central stem (×2000); m = 'mouth'; f = flagellae; l = leucosine inclusion; n = nucleus; p = chloroplast. After Lecal-Schlauder (1951, fig.45)

Fig. 7.2 *Coccoliths pelagicus* (Wallich, 1877): A, flagellate stage (*Crystallolithus hyalinus*) with visible haptonema (h) and small holococcoliths not arranged in a coccosphere (×1000); B, diagram of a holococcolith (×20 000); C, non-flagellate stage with placolith coccosphere (×1600); D, schematic section of a placolith: dd = distal disc; pd = proximal disc (×6000); E, arrangement of placoliths in the coccosphere (×2500). After various authors including Gaarder & Markali (1956, fig.1)

Biological Cycle

When conditions are favourable, each cell divides rapidly from one to five times per day. Division takes place within the cell and without loss of the flagellae. The shell is either abandoned or shared in equal parts by the two offspring cells, which subsequently reconstruct a complete shell. The coccoliths are formed in a vesicle within the cytoplasm after which they migrate to the periphery of the cell.

The life cycle of coccolithophores is complex and has remained obscure because of the great difficulties in obtaining unispecific cultures. What knowledge there is of the cycle is restricted to isolated observations.

The species *Coccolithus pelagicus* (Fig. 7.2) passes through at least two successive stages:

- The first, the *Crystallolithus hyalinus* stage, is perhaps haploid, flagellate and with tiny coccoliths (1.5 μm) called **holococcoliths**, which are not linked to form a coccosphere.
- The second, the *Coccolithus pelagicus* stage, is perhaps diploid (the sexual

stage), non-flagellate and with coccoliths of the placolith type (4.5 to 13 μm) arranged in a coccosphere.

Several other planktonic and benthic stages, both with and without coccoliths, have been observed, and although their significance has not been fully established, it is environmental, the benthic stage occurring in hostile environments.

In other species, the flagellate stage occurs with coccoliths, which may or may not be arranged in a coccosphere. Alternatively, coccoliths may be entirely absent.

Ecology

Coccolithophores are planktonic and are almost all marine and autotrophic. They inhabit waters that are poor in nutrients but rich in oxygen and sufficiently well lit to allow photosynthesis. Their maximum concentrations occur above a depth of 50 m in the tropics and above 10 to 20 m in temperate seas. Individuals that live at great depths (up to 4000 m) are probably heterotrophic and saprophytic.

Some species are established in cold waters but most prefer warm and temperate waters. On the whole, they are adapted to normal salinity and have an oceanic habitat although certain species flourish in coastal waters. Three species live exclusively in fresh water. Some can tolerate variations in salinity and can live in the hypohaline waters of estuaries. *Coccolithus fragilis*, the dominant species of Mediterranean plankton, has been found in the surface waters of the Dead Sea.

2. THE SYSTEMATIC CLASSIFICATION OF COCCOLITHS AND ITS PROBLEMS

Coccoliths are extremely small, measuring usually less than 15 μm and sometimes less than 1 μm. Their size and state of preservation in the fossil record make their systematic classification very difficult.

Minute Size

For a century, coccoliths were examined from rubbings under the light microscope. The characteristics discovered by using this instrument to the limit of its capabilities gave some idea of the morphology of coccoliths and made possible an initial classification. In this way, various groups were distinguished, such as:

- **Discoliths**: saucer-shaped, either circular or elliptical, sometimes with a stem rising from the base.
- **Lopadoliths**: vessel-shaped with a high raised edge.
- **Placoliths** or **tremaliths**: cuff-link shape with two discs linked by a median cylinder.
- **Rhabdoliths**: shield-shaped surmounted by a stem.

- **Zygoliths:** formed from an elliptical ring and a basket-handle arch or a transverse rectilinear arch.

Examination under polarized light revealed the structure of coccoliths and led to the distinction between the **helioliths** and the **ortholiths**, the former consisting of many crystalline elements with a radially disposed optical axis giving a black cross between crossed nicols, and the latter consisting of only a few crystals.

The recent use of electron microscopes to observe individuals isolated by rubbing, or *in situ* in nanofacies, has produced more precise data. It has been discovered, for example, that certain coccoliths are formed from the juxtaposition of very small crystals of uniform size (from 700 to 1000 Å) and form (rhombohedral or hexagonal prism). These **holococcoliths** are distinct from the **heterococcoliths**, which have crystals of different shapes and sizes.

The older classification was based on optical properties not all of which (e.g. birefringence) are visible in the electron microscope. This this has been added another classification based on more precise characteristics that are not optically visible (number and arrangement of constituent elements). For a time, the two classifications were a source of confusion but the distinction between them become less sharp as specialists have come to use both types of microscope for their observations.

Preservation and Distribution in Sediments

The miniscule and fragile calcareous shields are sensitive to the phenomenon of diagenesis. The holococcoliths disintegrate readily and do not fossilize well, the crystalline elements of the living organism being embedded in an organic matrix. They are found only in a few marly Palaeogene sediments. The heterococcoliths are more robust but, during dissolution or fragmentation, they may lose one or several of their elements: a discolith or a zygolith may appear as a simple circular ring (**cyclolith**) or an elliptical ring (**cricolith**). In other cases, re-crystallization during diagenesis may considerably modify the appearance of the coccolith (Fig. 7.18).

Although some sediments contain an abundance of coccospheres, coccoliths are usually dispersed on fossilization. Moreover, it is known that:

- a single coccosphere may bear different coccoliths; and
- a single species goes through different coccolith stages.

It follows, therefore, that a species cannot be determined with certainty from one isolated coccosphere.

For practical purposes, coccoliths are defined in terms of their form and structure and are given a binomial nomenclature. As a result of this classification, coccoliths that may perhaps derive from different coccolithophores can be assigned to a single 'genus'. To avoid any confusion between biological taxa and artificial morphological taxa, the use of a parataxonomy has been recommended. Each morphological type of coccolith is provisionally treated and named as an independent parataxon. In this way, the nomenclature can be unified later if there is found to be a correlation between the various fragments of the individual.

Description of Some Types

At present, there is no natural and coherent classification of coccolith fossils although several proposals have been made. These differ so much from one another that certain specialists have felt they would be better off classifying the different taxa by alphabetical order! However, several coherent groups can be isolated on the basis of the morphology of coccoliths found both on living individuals and/or on fossil coccospheres (Fig. 7.3–28):

- The Coccolithidae with placoliths (e.g. *Coccolithus*, *Cyclococcolithina*, *Helicopontosphaera*, *Watznaueria* and perhaps also *Prediscosphaera*).
- Calciosoleniaceae with scapholiths (e.g. *Anoplosolenia* and *Scapolithus*).
- Syracosphaeraceae with discoliths (e.g. *Syracosphaera*).
- Rhabdosphaeraceae with rhabdoliths (e.g. *Blackites*, *Rhabdosphaera* and *Sphenolithus*).
- Scyphosphaeraceae with lopadolith (e.g. *Scyphosphaera*).
- Arkhangelskiellaceae, consisting entirely of discolith fossils (Cretaceous) with a complex base, rarely found as coccospheres (e.g. *Gartnerago*).

Figs. 7.3–18 Calcareous nanofossils: coccoliths and coccospheres; p = profile; dv = distal view; pv = proximal view

3 *Zygrhablithus* (Eocene–Oligocene): holococcolith zygolith; from left to right, p (×2500), surface detail (×12 500) and dv (×2500)

4 *Cyclococcolithina* (Oligocene–Recent): placolith; from left to right, dv, pv and oblique p (×2500)

5 *Helicopontosphaera* (Eocene–Recent): placolith; pv (×2500)

6 *Watznaueria* (Carboniferous?, Jurassic–Cretaceous): placolith; from left to right, dv between crossed nicols (×5500), pv and oblique dv (×4500)

7 *Prediscosphaera* (late Cretaceous): from left to right, p (×1500), distal extremity of the stem (×7000) and dv with stem broken at the base (×2500)

8 *Scapholithus* (Cretaceous–Recent): dv of isolated scapholith (left) (×3200) and coccosphere of *Anoplosolenia* (right) (×500)

9 *Syracosphaera* (Oligocene–Recent): p of a 'buccal' discolith with central stem (left) and pv of base with rods and elliptical pores between (right) ×2500)

10 *Blackites* (Eocene–Oligocene): rhabdolith; (left) p and dv (right) (below) (×9000)

11 *Rhabdosphaera* (Pliocene–Recent): rhabdolith coccosphere (×1500)

12 *Sphenolithus* (Eocene–Pliocene): p (left), (×2000) and pv (right) (×1700)

13 *Scyphosphaera* (Miocene–Recent): lopadolith; from left to right, p, pv (×2000) and coccosphere (×500)

14 *Neococcolithes* (= *Zygolithus*) (Cretaceous–Eocene): zygolith; p (left) and dv (right) (×1750)

15 *Gartnerago* (late Cretaceous): discolith with base formed from four areas of large joined elements; from left to right, dv, p and oblique pv (×2800)

16 *Eiffellithus* (late Cretaceous): from left to right, p (above), pv (below) and dv (×2000)

17 *Crucirhabdus* (Triassic? Lias): from left to right, dv, p and section (×3000)

18 *Stephanolithion* (Jurassic–Cretaceous); above and from left to right, dv, section p and two specimens with diagenetic crystalline growth seen under the light microscope (approx ×2000); below, reconstructed oblique pv (left) and dv (right) (×12 000)

After photographs and drawings by Buckry, Deflandre, Farinacci, Gartner, Kamptner, Lezaud, Noel, Perch-Nielsen, Prins, Roth, Samtleben, Zeighampour, etc.

CALCAREOUS NANOFOSSILS

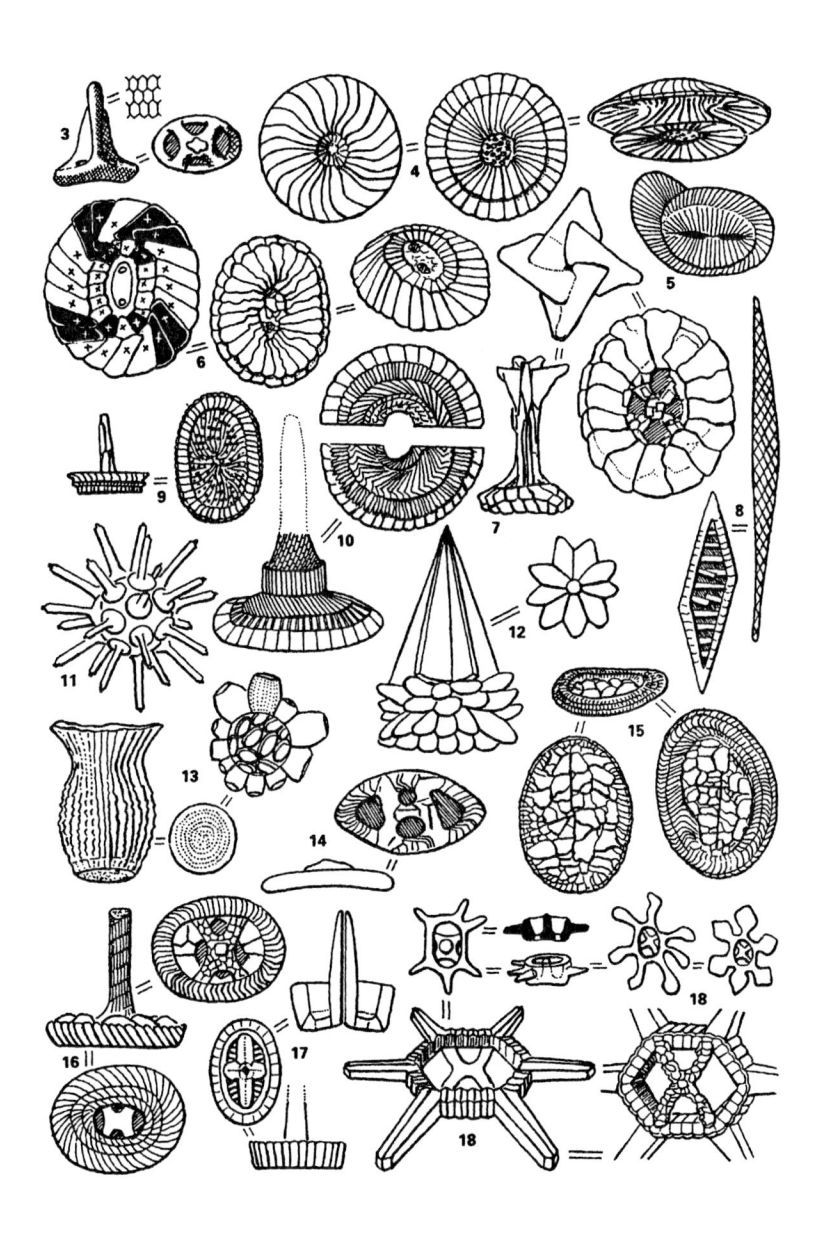

Although they have not been found in the form of coccospheres, asteroliths characterize the Discoasteridae group, to which might also be added *Fasciculithus, Heliolithus* as well as perhaps *Braarudosphaera* for which coccospheres are known.

These 'families' by no means contain all the coccoliths in the 200 'genera' and some 2000 'species' that are known. An example of this is provided by the zygoliths, which are of holococcolith type in *Zygrhablithus* and of heterococcolith type in *Neococcolithes, Eiffelithus, Crucirhabdus* and *Stephanolithion*.

Finally, there are other nanofossils which, despite an unusual appearance, are considered coccoliths. It is possible that some of these (*Marthasterites, Isthmolithus* and *Lithostromation*) may, in fact, be intracytoplasmic calcareous structures situated within the coccosphere, as has been proved for *Ceratolithus*.

3. COCCOLITHS THROUGHOUT GEOLOGICAL TIME

Emergence and Spread

The almost total absence of calcareous nanofossils in the Triassic and the Palaeozoic is hardly surprising as the sediments of these periods, often strongly consolidated, do not lend themselves readily to the preparation of rubbings. This absence, however, is undoubtedly more apparent than real. A few restricted observations suggest that there are surprises still in store awaiting methodical examination of Triassic and Palaeozoic nanofacies.

The oldest known coccoliths date from the end of the Palaeozoic but the

---

Figs. 7.19–36 Discoasterid calcareous nanofossils (19–21) and other presumed coccoliths (22–28), microrhabdulids (29–32), calcareous dinocysts (33), *Thoracosphaera* (34), *Schizosphaerella* (35) and *Nannoconus* (36)

19 *Heliodiscoaster* (Palaeocene–Miocene): opposite sides (left) and oblique view of another individual with stem (right) (×2000)
20 *Turbodiscoaster* (Palaeocene–Oligocene): seen from in front under the light microscope (left) (×1000) and oblique view in SEM (right) (×1600)
21 *Eudiscoaster* (Miocene–Pliocene) (×3200)
22 *Fasciculithus* (Palaeocene): from left to right, frontal, lateral and sectional views (×3000)
23 *Heliolithus* (Palaeocene); from left to right, frontal, lateral and sectional views (×1200)
24 *Braarudosphaera* (Cretaceous–Recent); from left to right, coccosphere viewed laterally (above) and frontally (below) (approx. ×1500)
25 *Marthasterites* (Cretaceous–Eocene) (×1500)
26 *Isthmolithus* (Eocene): front view (above), lateral view (below) (×1500)
27 *Lithostromation* (Eocene): opposite sides (×1500) and lateral view (centre) (×500)
28 *Ceratolithus* (Miocene–Recent): different views of a specimen (×1000; ×2000 above)
29 *Microhabdulus* (late Cretaceous) (×1500); diagrams (left) (×3000)
30 *Lucianorhabdus* (late Cretaceous) (×1500)

CALCAREOUS NANOFOSSILS

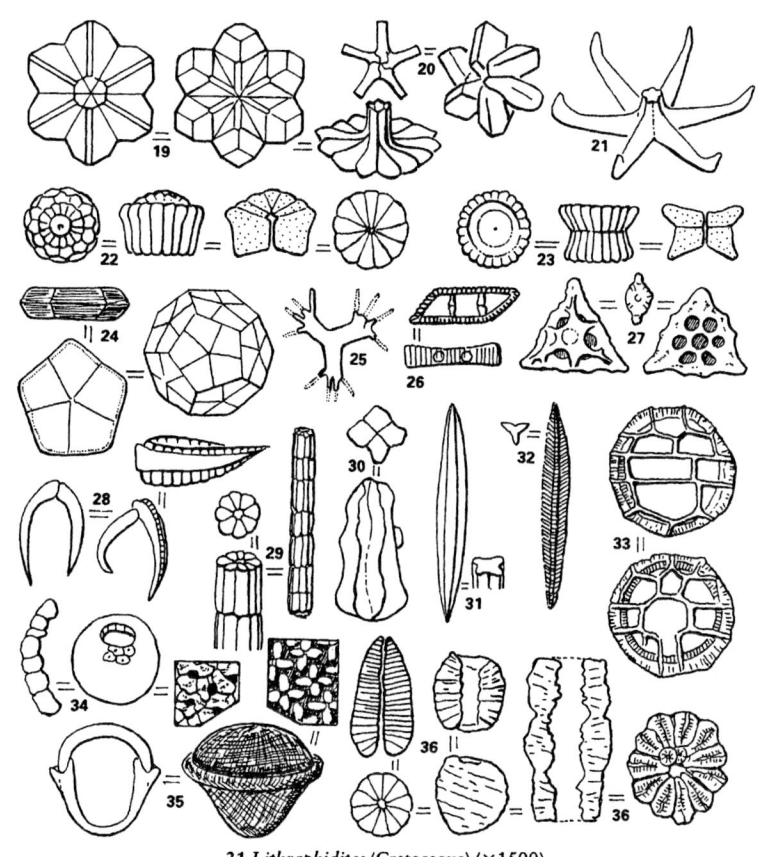

31 *Lithraphidites* (Cretaceous) (×1500)
32 *Triquetrorhabdulus* (Oligocene–Miocene) (×2500)
33 Calcareous dinoflagellates of calciodinellid (*Calciodinellum*): dorsal (above) and ventral sides (below) (×700)
34 *Thoracosphaera* (Cretaceous–Recent): from left to right, section of wall (×2000), overall view (×850) and detail of surface with joined crystals (×3100)
35 *Schizosphaerella* (Triassic–early Cretaceous): from left to right, section, overall view (×600) and detail of surface showing juxtaposition of crystals arranged in a cross (×8000)
36 *Nannoconus* (Triassic, Jurassic, late Cretaceous): different species (×1350); rosette-shaped assemblage (right) (×400)

After drawings and photographs by Aubry, Black, Bronnimann, Deflandre, Dépêche, Farinacci, Fütterer, Hay, Kamptner, Martini, Noel, Perch-Nielsen, Prins, Trejo, etc.

Fig. 7.37 Stratigraphic ranges of some groups of calcareous nanofossils

group did not become common until the Lias (*Crucirhabdus*). At the end of the Jurassic (e.g. *Stephanolithion* and *Watznaueria*) and especially during the Cretaceous, there was a multiplication of complex forms, the accumulations of which gave rise to the chalks. These coccoliths disappeared at the end of the Cretaceous. Few 'genera' are represented in the Mesozoic and the Cenozoic, the latter era being especially marked by the development of the Discoasteridae. This group evolved over the 50 million years from the Palaeocene to the end of the Pliocene: rosette-shaped forms (*Heliodiscoaster*) were succeeded by others with a few slender arms (*Eudiscoaster*).

Palaeoecology

Fossil coccoliths are mostly found in marine series deposited far from coasts but they sometimes occur in littoral sediments. Certain associations lived in cold waters and others (e.g. discoasterids and *Sphenolithus*) in relatively warm environments. No coccoliths have been found in lacustrine sediments. Although some have been indicated in closed basin formations, it cannot be asserted that these were lagoonal. The very thin (a few millimetres) beds of coccolith chalk which, in the Limagne Oligocene, alternate with '*Cypris*' limestones seem to correspond to brief episodes during which lakes and lagoons communicated with the open sea.

4. SOME OTHER CALCAREOUS NANOFOSSILS

The nanofossils that have been mentioned are all related with some degree of certainty or probability to the Coccolithophyceae. Others, however, have different and more obscure affinities.

The microrhabulids (Figs 7.29–32), which are known from the Cretaceous, take the form of a pyriform or acicular cylindrical rod.

The different *Nannoconus* species (Fig. 7.36) are cylindrical globular or pyriform assemblages (length = 15 to 50 μm, width = 5 to 15 μm). The tiny triangular plates are arranged in a whorl with their points directed towards the axis of the assemblage where there is a channel of varying width open at both ends. *N. multicadus* forms articulated chains. A score of species succeeded each other from the end of the Jurassic to the end of the Cretaceous in the regions bordering the Mediterranean and the Caribbean. They abound in the fine limestones (e.g. chalks and 'majolica') and marls together with coccoliths, radiolaria, calpionellids and planktonic foraminifera. The organism that produced *Nannoconus*, however, remains unknown. However, the arrangement into a rosette shape around a tiny central cavity suggests that it could be a coccolithophore.

Certain dinoflagellates in the class Dinophyceae – the calciodinellids – secrete spherical calcareous cysts (dinocysts) with a diameter of 25 to 45 μm (Fig. 7.33). Like the sporopollenin forms that will be discussed later (Chapter 10), these cysts open via an archaeopyle and an operculum. They are ornamented with spines or with ridges that emphasize the tabulation. The wall consists of a single layer of irregular crystals with pores between them.

CALCAREOUS NANOFOSSILS 81

Calciodinellids were found in the fossil state before living examples were discovered.

Although non-tabular, the spherical shells of *Thoracosphaera* (Fig. 7.34) include an archaeopyle and operculum and are also grouped with the calcareous dinocysts. They are abundant in certain levels such as the Danian (early Palaeocene).

Mention must also be made of forms called *Schizosphaerella* (Fig. 7.35), which are very common in Jurassic marine sediments. They are shells with a diameter of 5 to 30 μm linked together as two dissimilar valves. The wall has a remarkable structure with small tabular crystals arranged regularly in a cubic lattice. Their systematic position remains unknown.

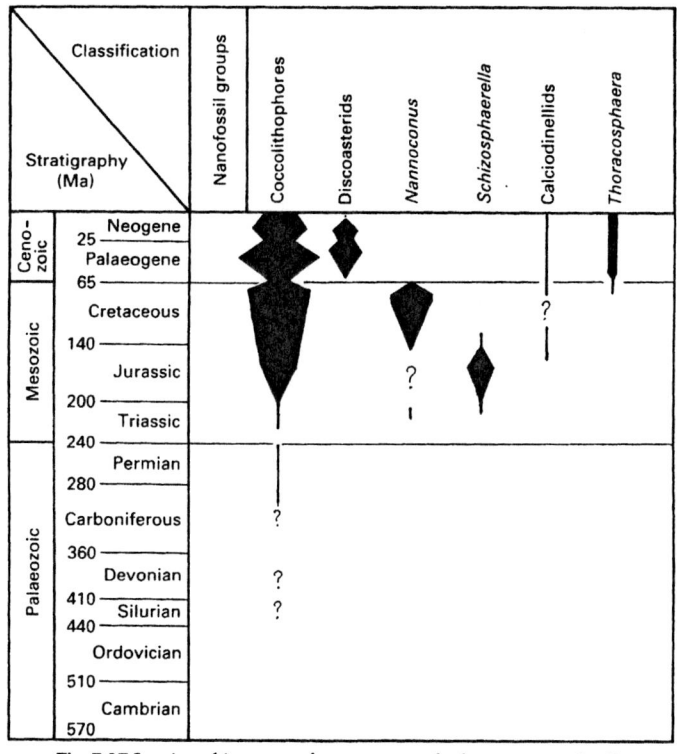

Fig. 7.37 Stratigraphic ranges of some groups of calcareous nanofossils

CONCLUSION

Most calcareous nanofossils are the remains of marine planktonic protophytes. Virtually no evidence of them has been found in lacustrine sediments but this may simply be because nobody has looked for them there. Their role in petrogenesis is very important. Moreover, although some species span long periods most existed for only a brief time and so are excellent stratigraphic markers (Fig. 7.37). The value of coccoliths to geology – especially in the reconstruction of palaeoenvironments – would be still greater and would carry more weight if more were known about the biology of extant Coccolithophyceae.

BIBLIOGRAPHY

The beginner may obtain an overview of the coccolithophores by reading W.W. Hay, Calcareous nanofossils, in A.T.S. Ramsay (ed.), *Oceanic Micropalaeontology* (vol. 2, pp. 1055–1200, 1977), and the chapter devoted to the subject in H. Tappan, *The Paleobiology of Plant Protists*, pp. 678–803 (W.H. Freeman, San Francisco, 1980).

The following monographs may be referred to for a more specialized approach: D. Noel, *Sur les Coccolithes du Jurassique européen et d'Afrique du Nord* (CNRS, Paris, 1965); B. Prins, 'Evolution and stratigraphy of coccolithinids from the lower and middle Lias', in *Proceedings of the First Planktonic Conference (Geneva, 1967)*, vol. 2, pp. 547–558 (1969); D. Noel, *Coccolithes crétacés* (CNRS, Paris, 1970); K. Perch-Nielsen, 'Durchsicht Tertiärer Coccolithen', in *Proceedings of the Second Planktonic Conference (Rome, 1970)*, vol. 2, pp. 939–980 (1971); B. Prins, 'Speculations on relations, evolution and stratigraphic distribution of discoasters', ibid., pp. 1017–1038.

Two catalogues are now in the process of publication: the first by A.R. Loeblich and H. Tappan, which appeared in *Journal of Palaeontology*, 47 (4) 715–759 (1973), and continues in *Newsletter of the International Nanoplankton Association* (2 issues per year); and the second by A. Farinacci, *Catalogue of Calcareous Nannofossils* (Techno-Scienza, Rome, 1969–; 8 vols have appeared).

Chapter 8
Siliceous Microfossils

The siliceous microfossils derive initially from fragile opaline skeletons, which may dissolve completely before or after fossilization. Alternatively, during diagenesis, they may undergo a process of epigenesis, developing into 'ghost' forms of quartz, calcite, dolomite, pyrite, phosphate, etc. Moreover, they are often contained in very compact rocks. For all these reasons, they are difficult to study and our knowledge of many fossil species remains incomplete, especially from pre-Cenozoic sediments.

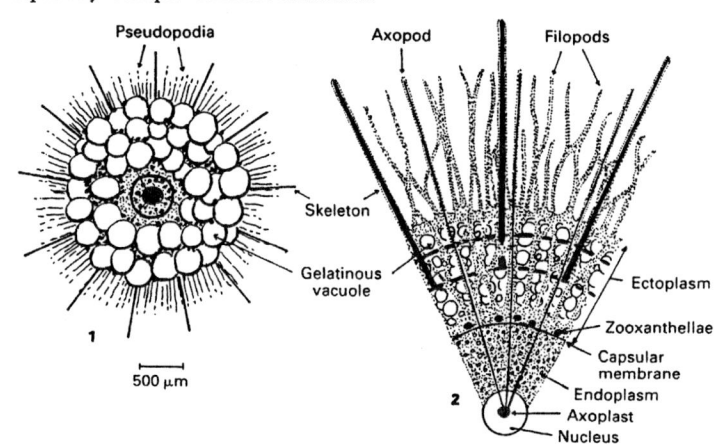

Fig. 8.1 External appearance and optical section of a living spumellarian polycystine radiolarian ($\times 14$)

Fig. 8.2 Diagram of part of a section of a spumellarian, the skeleton of which is formed from two concentric shells, trabeculae and radial spines

1. POLYCYSTINE RADIOLARIANS

Living Radiolarians

These rhizopod protozoa are characterized by a rigid 'pseudochitinous' **capsular membrane** of unknown chemical composition. It divides the organism into two parts: a vacuolar **ectoplasm**, which secretes the skeleton or test and houses symbiotic zooxanthellae, and an **endoplasm**, which contains the nucleus and a variety of inclusions (Figs 8.1 and 2). The **pseudopodia** are of two types:

- Filipodia, thin and ramified, simple extensions of the ectoplasm.
- Axopodia are never ramified and have a rigid axis converging on an axoplast that may be juxtanuclear or intranuclear.

As radiolarians cannot be cultured in the laboratory little is known about them. They may live for about a month. They feed on small living prey such as diatoms and copepods. Multiplication takes place by binary cell division. One of the offspring keeps the original skeleton while the second produces a new one with centrifugal growth. Intraspecific dimorphism has been interpreted as

---

Fig. 8.3–21 Different types of polycystine radiolarians: Albaillellaria (3), Spumellaria (4–15) and Nassellaria (16–21)

3 *Albaillella* (Carboniferous); latticed test around spicule (×350)
4 *Haplentactinia* (Devonian): entactinid with central spicule and spines traversing a slightly eccentric spherical shell (×130)
5 *Collosphaera* (Miocene–Recent): thick shell with relatively few pores (×250)
6 *Cenosphaera* (Eocene–Recent): single shell without spines (×100)
7 *Actinomma* (Miocene–Recent): several concentric shells with radial spines (×235)
8 *Saturnalis* (Pleistocene–Recent): two polar arms joining to form a concentric ring (×135)
9 *Ommatartus* (Miocene–Recent): elliptical shell with five constrictions (×250)
10 *Cannartus* (Oligocene–Miocene): shell with two lobes and two hollow polar arms (×135)
11 *Capnuchosphaera* (Triassic): detail of a radial appendage (×200)
12 *Alievium* (Cretaceous): spongy disc with three marginal spines (×85); on the left, transverse section (×185)
13 *Amphirhopalum* (Pliocene–Recent): numerous concentric shells with several arms (×150)
14 *Paronaella* (Carboniferous–Cretaceous): three spongy arms with pores arranged in rectilinear lines (×45); on the right, detail (×180)
15 *Spongaster* (Miocene–Recent): spongy disc with four arms linked together by a spongy sheet (×45); on the right, detail (×170)
16 *Zygocircus* (Eocene–Recent): single cephalis (×200)
17 *Giraffospyris* (Eocene–Recent): single lobed cephalis (×150)
18 *Dorcadospyris* (Eocene–Miocene): rear view of a complete individual (×100); view from below (left); front view with cephalic constriction (right) (×150)
19 *Callimitra* (Eocene–Recent): cephalis with spicule visible (×270)
20 *Corocalyptra* (Pleistocene–Recent): cephalis with visible spicule and thorax (×150)
21 *Botryopyle* (Eocene–Recent): multilobed cephalis (×300)

After de Wever, Foreman, Goll, Holdsworth, Nigrini, Pessagno, Riedel, etc.

SILICEOUS MICROFOSSILS

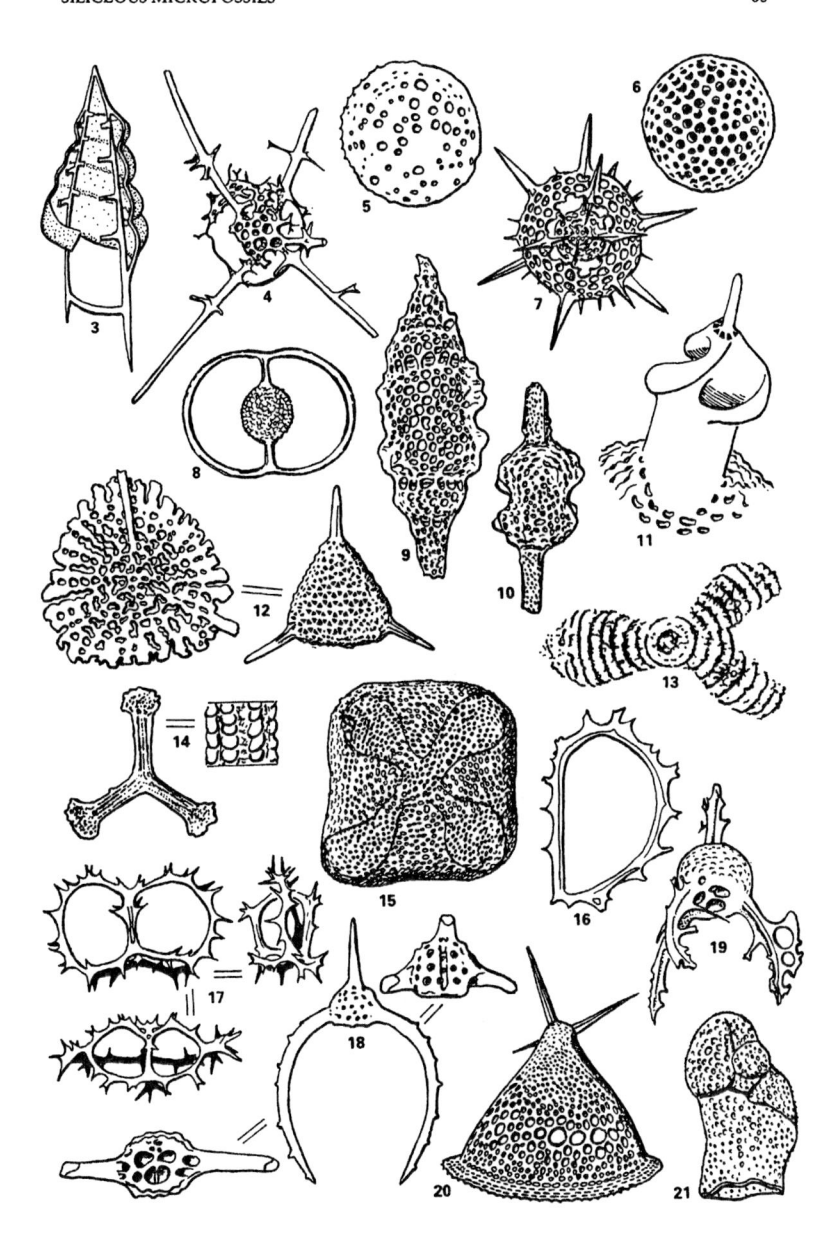

85

indicating two successive stages of a cycle, perhaps analogous to that of foraminifera.

All radiolarians are marine, stenohaline (salinity >30 parts per mille ($^o/_{oo}$)) and planktonic. At a depth of 100 m, they are found in great number and variety. Towards the surface and below a depth of 500 m, however, the density and diversity decline. As a general rule, the tests of individuals found in surface waters are smaller and thinner than those living at depth. Finally their size is in inverse proportion to thickness, which diminishes from the equator towards the poles.

Skeleton

The skeleton or test, which is the only part to be preserved, may not be present during the lifetime of the animal. In most cases, it consists of spicules that are either isolated or arranged in complex structures. They range in size from 100 to 400 µm and even, in exceptional cases, to 2 mm long and are formed from almost pure opal. They appear either as spongy masses or trabeculae forming latticed surfaces.

The geometry of the skeleton and distribution of perforations on the capsular membrane are used to subdivide the living Radiolaria into two main groups – the orders **Spumellaria** and **Nassellaria**, which are now usually united in the superorder Polycystina. The Spumellaria (Figs 8.4–15) have a radial symmetry derived from a spherical shell. Variations relate to:

- The form of the shell (e.g. spherical, elliptical, lenticular or cubic), their number (i.e. single or colonial, concentric or linked by trabeculae) and their structure (spongy or latticed).
- The form (ramified or not), number, arrangement, relative size and structure of the radial extensions.

The Nassellaria (Figs. 8.16–34) have in common a latticed test resulting from the modification of a primary **spicule** and this is in the form of a bar, the ends of which bear bundles of **spines**. Sometimes visible, this spicule is included in a small shell or **cephalis**, which varies in both size and form: spherical or lobed, with few or many perforations, with or without spines or tube. The cephalis

---

Fig. 8.22–34 Nassellarian polycystine radiolarians (contd): a = abdomen; c = cephalis; pa = post-abdomen; t = thorax

22 *Anthocyrtidium* (Oligocene–Recent): cephalic constriction in form of diagonal furrow (f) (×250)
23 *Phormostichoastus* (Oligocene–Recent): cephalis with oblique apical tube (×230)
24 *Theocampe* (Cretaceous–Recent): left, axial section showing the diagonal apical tube (at); right, apical view of cephalis (×300)
25 *Holocryptocapsa* (Cretaceous): axial section showing the cephalis and the thorax included in the abdomen (×230)
26 *Peripyramis* (Oligocene–Recent): cephalis much reduced (×150)
27 *Pterocanium* (Eocene–Recent) (×250)
28 *Calocyclas* (Eocene) (×150)
29 *Lychnocanoma* (Eocene–Recent) (×250)

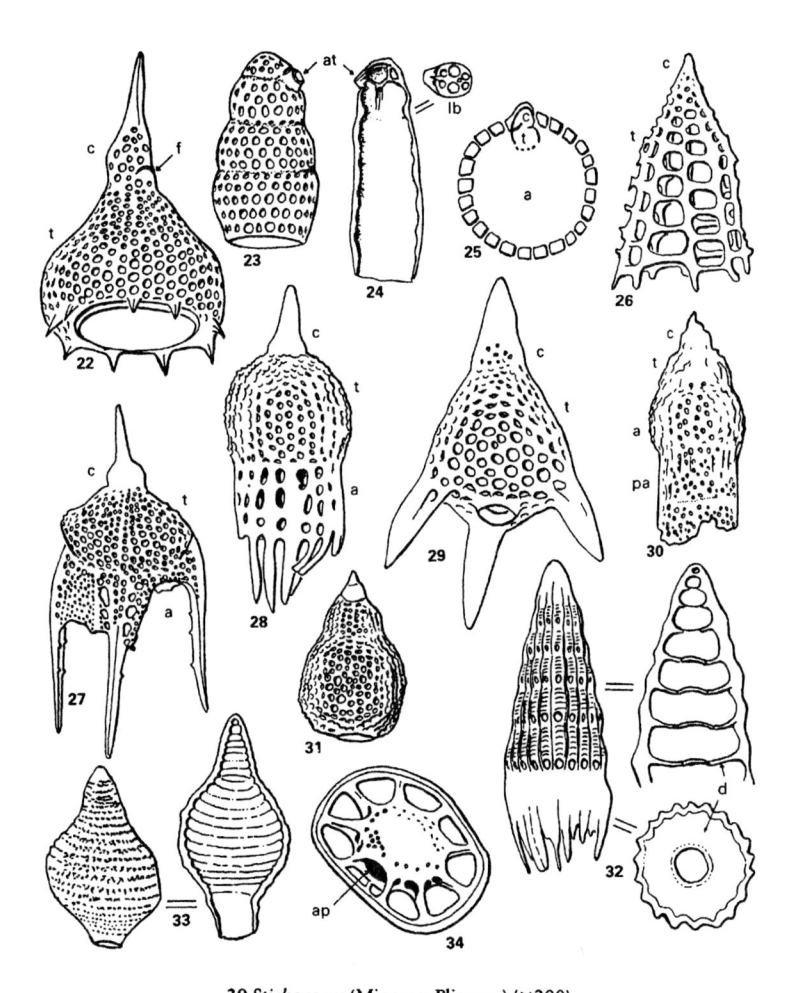

30 *Stichocorys* (Miocene–Pliocene) (×200)
31 *Cyrtocapsa* (Oligocene–Miocene) (×140)
32 *Dictyomitra* (Cretaceous): right and above, axial section (d = pierced septum); right and below, transverse section at the level of a diaphragm (×180)
33 *Mirifusus* (Jurassic–Cretaceous): numerous postabdominal chambers separated by septa (×70)
34 *Saturniforma* (Cretaceous): lenticular thorax with aperture (ap) and radial spines forming a concentric ring (×200)

After Foreman, Nigrini, Pessagno and Riedel.

alone may form the test. Equally, it may be extended by uniserial chambers (termed thorax, abdomen and postabdomen), which may or may not be separated by septa with pores. The resulting structure exhibits an almost axial symmetry.

These two orders are used to classify the Cenozoic and Mesozoic polycystines but those dating from the Palaeozoic are less easily labelled. If the spherical entactinids (Fig. 8.4) fit without great difficulty into the Spumellaria, the Albaillellaria (Fig. 8.3) seem to bear no relation to either group.

Even though some 1200 genera and 7000 species have been described, fossil radiolarians are still imperfectly known and since the abandonment of Haeckel's (1887) artificial classification a full-scale revision of the system is under way.

Radiolarians Throughout Geological Time

The Radiolaria appear in the Ordovician (Arenig). During the Palaeozoic, the entactinids predominated in association with the Albaillellaria and true Spumellaria, the latter having a latticed shell or, from the Devonian, a spongy one.

The beginning of the Mesozoic witnessed great changes. The characteristic forms of the Palaeozoic disappeared while the first Nassellaria emerged, a group which became progressively more important. The Spumellaria remained dominant, being represented by several particular genera (*Capnuchosphaera* in the Triassic, *Paronaella* in the Jurassic–Cretaceous, *Alievium* and rotaformids in the Cretaceous).

In the Cenozoic, many forms disappeared while new ones emerged which, with few exceptions, have remained up to the present. The diversification of the polycystines was at its height. The Spumellaria gradually lost ground to the Nassellaria. From the Neogene onwards, the tests become, on the whole, thinner, a fact which some authors have interpreted as resulting from increased competition from diatoms.

Radiolarians of the Cenozoic had a mode of life identical to that of living forms. The same is true for the Mesozoic species whether associated with planktonic organisms (e.g. ammonites, foraminifera, calpionellids, *Nannoconus* and coccoliths) or with nektonic ones (conodonts in the Triassic). Nevertheless, the fact that they are also found with corals, rudist bivalves and calcareous algae, suggests that some deposits must have been laid down in shallow marine environments close to the continent.

As far as the Palaeozoic is concerned, conclusions are more tentative. Although radiolarians are often associated with goniatites and graptolites, they are as frequently found with trilobites, lingulid brachiopods, ostracods, and even with fragments of continental plants (*Lepidodendron*). Lastly, it has been suggested that the Albaillellaria, being ill-suited to floating, formed part of the coastal benthos.

2. DIATOMS

Living Diatoms

Diatoms (class Bacillariophyceae) are autotrophic unicellular chrysophyte algae with large olive-green or brown chloroplasts. The cell many be as large as 2 mm but on average ranges from 40 to 50 µm in diameter. It is supported by the **frustule**, an opaline intracytoplasmic skeleton (Fig. 8.35).

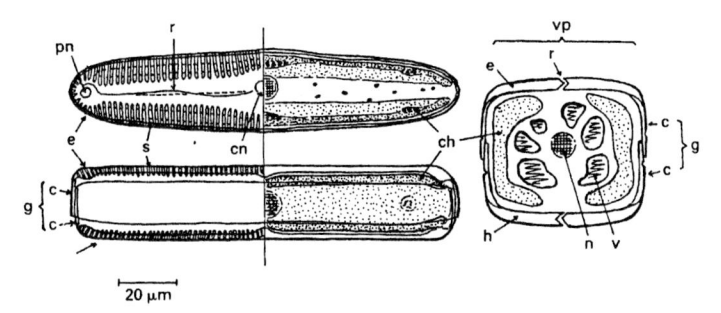

Fig. 8.35 Diagram of a living euraphid pennate diatom (*Pinnularia*). On the left, (above) valve view, (below) connective view with the left half showing the appearance of the frustule and the right half the content of the cell (approx. ×525); on the right, transverse section (approx. ×1040); c = cingulum; g = girdle; ch = chloroplast with pyrenoids; e = epitheca; h = hypotheca; n = nucleus; cn = central nodule; pn = polar nodule; r = raphe; s = striae (lines of pores separated by costae); v = vacuole; vp = valve plate. After Pfitzer (1974, fig. 90)

The frustule consists of two **valves** or **theca** which differ slightly in size. Both have a **valve plate** and **cingulum**. The smaller of the valves, the **hypotheca**, fits into the larger one or **epitheca** in such a way that one cingulum overlaps the other, forming a connective band or **girdle**.

When a cell divides, each of the new cells takes as its epitheca one of the valves of the parent frustule and within 10 to 20 minutes builds up its own hypotheca. Among benthic forms, the process of fission may take place three times a day while among planktonic forms the rate may be as high as eight times per day. The operation of this mechanism, however, progressively reduces the size of the frustule (Fig. 8.36) and, at a certain threshold, sexual reproduction comes into play with the formation of **auxospores** that restore individuals to the normal dimensions (fig. 8.37). Certain pelagic diatoms in the order Centrales (*Chaetoceras*) produce resistant cysts (**endospores** or **statospores**) with a condensed cytoplasm. the bivalved shell is thick, devoid of girdle, and with an ornamentation different from that of the frustules (Figs. 8.38 and 39). After a certain time, the endospores give rise to normal vegetative (i.e. asexual stage) cells.

Diatoms live either in isolation or linked together in colonies by their appendages or a mucous segment. Although numerous genera are found in all types of aquatic environment regardless of salinity (fresh water, salt water) and

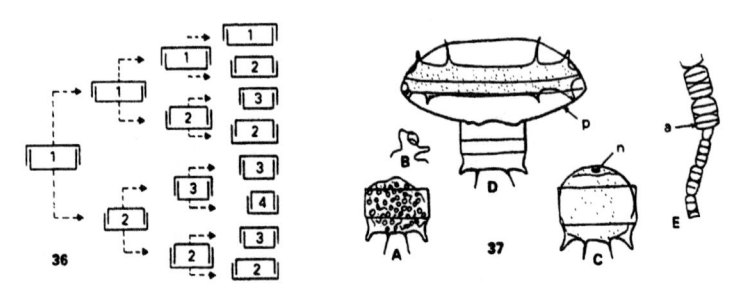

Fig 8.36 Asexual reproduction in diatoms. In *Coscinodiscus*, the initial cells, measuring 200 μm, produce cells the size of which diminishes by between 1 and 2 μm with each division. The regenerated valve is always the hypotheca. Mitosis ceases when the cells reach 55 to 60 μm

Fig. 8.37 Sexual reproduction in the diatom *Biddulphia*: A, production of antherozoids in a valve of the parent frustule (×200); B, biflagellate antherozoid (×540); C, oosphere in a valve of the parent frustule (n = nucleus) (×200); D, construction of the frustule of the auxospore within the perizonium (p), a membrane limiting the maximum extension of the oosphere after its departure from the parent frustule and fertilization (×200); E, row of cells showing abrupt increase in size after the appearance of the auxospore (a) (×80).

After Bergon (1974, figs 133 and 138)

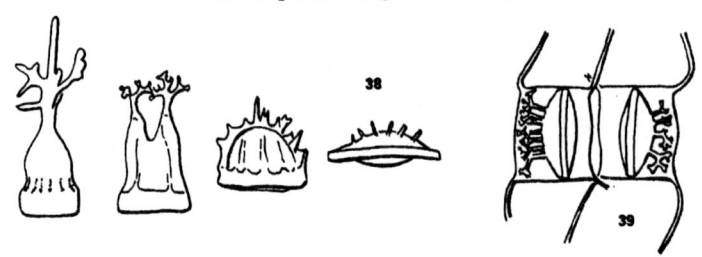

Fig. 8.38 Separate and connected (right) valves of fossil cysts. Although attributable to different species of the centric diatom *Chaetoceros*, each has been given particular names of species and genera (×1180)

Fig. 8.39 Living cells of *Chaetoceras* with cysts (×360). After Gran (1933, fig. 96)

temperature (from the polar icecaps to hot thermal springs), most species are subject to strict ecological constraints. Many varieties are benthic and these may be fixed or mobile. Others have a planktonic habitat in both fresh water (particularly the Pennales) and the sea (particularly the Centrales). Diatoms are especially abundant in the plankton of the Antarctic seas.

Frustule

The frustule represents some 15 to 20% of the total dry weight of the cell although the figure may be as low as 1% in some only slightly silicified species and as high as 47% in *Coscinodiscus*. The form of the frustule varies from one species to another but all derive from the simple schema that has been outlined. The hypotheca and epitheca may or may not be identical. The girdle may consist of one or several bands. The valve plates are sometimes formed from a single sheet ornamented with costae and perforated by pores or **punctae**. In most cases, however, they are bilamellar (Fig. 8.40).

The two internal and external **lamina** are superposed on each other and are linked by a network of perpendicular partitions that bound the hexagonal **alveoli**. The floor and ceiling of the alveoli are penetrated by pores, which may or may not have **sieve plates**. Communication between contiguous alveoli is made possible by the presence of **windows**. The interior of the frustule is sometimes divided by **diaphragms**. Details such as these can be extremely small. *Pleurosigma angulatum* Smith 1853, which was once used to test the lenses of light microscopes, has slit external pores which are only 0.05 μm (= 500 Å) in width.

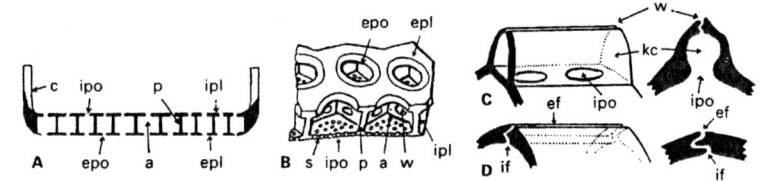

Fig. 8.40 Structure of the diatom frustule: A, longitudinal section of a valve; B, organization of a bilamellar valve plate (approx. ×5100); C, raphe of an archaeoraphid; D, raphe of an euraphid; a = alveolus; c = cingulum; ef = external fissure; epl = external lamina; epo = external pore; if = internal fissure; ipl = internal lamina; ipo = internal pore; kc = keel channel; p = partition; s = sieve; w = window

The classification of diatoms is based on the morphology of the frustule. Its form and the arrangement of the pores on the valve plates allow two main orders to be distinguished – the **Centrales** or centric diatoms and the **Pennales** or pennate diatoms.

The Centrales (Figs 8.41–52) have round or polygonal valves; the pores are in radiating, concentric rows; the overall symmetry is radial.

The Pennales (Figs 8.53–68) have elliptical valves and a bilateral arrangement of the pores. Many Pennales exhibit a distinctive morphological feature – a median or marginal **raphe** (Fig. 8.40) formed from the modification of a line of large coalescent alveoli. In the suborder Archaeoraphidina, the raphe opens on the inside via separate pores and on the outside through an elongated slit situated at the summit of a keel that runs uninterruptedly from pole to pole of the valve (Naviculales) or encircling it (Surirellales). The raphe of the Euraphidina is a simple fissure in the valve with a slit on either side. The Chamaeraphidina have an extremely small divided raphe on each valve. Among the Monoraphidina, only the hypotheca has a raphe; the epitheca is without, its place being occupied by a hyaline axial area called the **pseudoraphe**, which is neither fissured nor perforate. A final group, the Araphidina, have a pseudoraphe on each valve.

This classification is difficult to apply to certain planktonic forms.

92   PART 1: MICROFOSSIL GROUPS

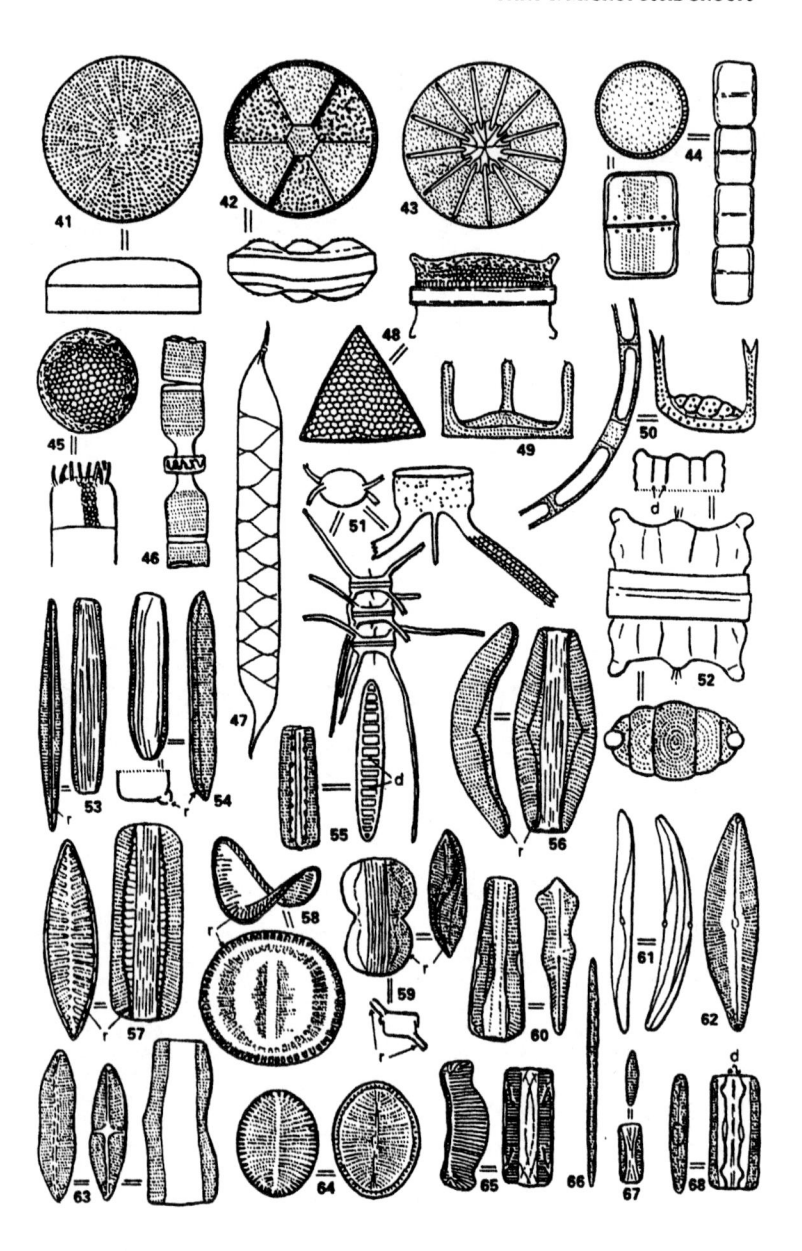

Figs 8.41–68 Centrale diatoms (41–52), archaeoraphid Pennales (53–58), euraphids (59–62), monoraphids (63–64), chamaeraphids (65) and araphids (66–68); l = lacustrine; m = marine; b = benthic; p = planktonic; cv = connective; vv = valve view; d = diaphragm; r = raphe

41 *Coscinoconus* (Cretaceous–Recent; l/m,p): from top to bottom, vv and cv (×250)
42 *Actinoptychus* (Cretaceous–Recent; m,p): from top to bottom, vv and cv; valve with triangular compartments alternately projecting or depressed (×200)
43 *Asterolampra* (Eocene–Recent; m,p): vv with hyaline central area and radial furrows (×400)
44 *Melosira* (Cretaceous–Recent; l/m,p): from left to right, vv (×600), cv (×450) and row of cells forming a colony (×300)
45 *Stephanopyxis* (Cretaceous–Recent; m,p): from top to bottom, vv (×650) and cv (×400)
46 *Strangulotrema* (Eocene; m,p): cv with two connected frustules (×540)
47 *Rhizosolenia* (Oligocene–Recent; l/m, p): cv, long cylindrical frustule with large girdle formed from bands (partly encircling) the valve (×230)
48 *Triceratium* (Cretaceous–Recent; m, b/p): from top to bottom, cv and vv (×200)
49 *Trinacria* (Cretaceous–Oligocene; m,p): cv of a valve (×200)
50 *Hemiaulus* (Cretaceous–Recent; m,p): from left to right, row of colonial cells (×60) and diagonal view of a valve (×270)
51 *Chaetoceros* (Miocene–Recent; m,p): from left to right, vv, colonial form (×130); cv of a valve (×260); elliptical valve with very long appendages. See also Figs 8.38 and 8.39
52 *Biddulphia* (Cretaceous–Recent; m, b/p): from top to bottom, section of a valve, cv and vv (×400)
53 *Bacillaria* (Oligocene–Recent; m, b/p?): from left to right, vv, cv and median raphe (×280)
54 *Nitzschia* (Miocene–Recent; l/m, b/p): from left to right, section of valve, cv, vv and marginal raphe (×280)
55 *Denticula* (Miocene–Recent; l/m, b/p): from left to right, cv and internal view of valve (×600)
56 *Epithemia* (Pliocene–Recent; l/m, b): from left to right, vv, cv and submarginal raphe (×160)
57 *Surirella* (Oligocene–Recent; l/m, b): from left to right, vv, cv and marginal raphe encircling the valve plate (×200)
58 *Campylodiscus* (Miocene–Recent; l/m, b): from left to right, profile view, vv and marginal raphe (×100)
59 *Amphiprora* (Miocene–Recent; l/m, b): from left to right, cv, section of valve, vv and marginal valve, sigmoid in shape, at summit of a keel (×180)
60 *Gomphonema* (Pliocene–Recent; l/m, b): from left to right, cv and vv (×300)
61 *Pleurosigma* (Miocene–Recent; l/m, b): from left to right, vv, cv and sigmoid median raphe (×200)
62 *Cymbella* (Miocene–Recent; l,b): vv (×350)
63 *Achnanthes* (Miocene–Recent; l/m, fixed b): from left to right, vv fixed hypotheca with raphe, vv epitheca with pseudoraphe and cv (×150)
64 *Cocconeis* (Oligocene–Recent); l/m, fixed b): from left to right, vv epitheca with pseudoraphe and vv fixed hypotheca with raphe (×360)
65 *Eunotia* (Miocene–Recent; l, b): vv and cv, short raphe, separated into two sections extending from polar nodules (×400)
66 *Synedra* (Eocene–Recent; l/m, b/p): vv (×100)
67 *Fragilaria* (Eocene–Recent; l/m, b/p?): from top to bottom, vv and cv (×50)
68 *Grammatophora* (Miocene–Recent; m,b): from left to right, vv and cv, corrugated and perforated internal diaphragm (×350)

After Bourelly, van Heurck and Wornardt.

Specialists in diatoms are, therefore, attempting, with the aid of the SEM, to establish a new systematic classification for the 200 genera and some 20 000 species, both fossil and extant, that have so far been described.

Diatoms Throughout Geological Time

Fossil diatoms are known from their frustules and endospores, the latter sometimes being more frequent than the frustules of the cells from which they derive. If it is accepted that the oldest frustules of *Chaetoceras* date from the Miocene, it is probably to this genus that various Palaeogene and Cretaceous cysts must be attributed although they bear different names.

The first indisputable diatoms date from the early Cretaceous (Barremian and/or Aptian). They became common, however, only after the Turonian. Some 70 genera and 300 species of Centrales have been found in marine Cretaceous deposits. This abundance suggests that diatoms must have appeared at an earlier date even though no traces are in evidence.

The first Pennales (*Pinnularia*) appear in the Eocene, which also saw the beginning of the movement of diatoms into fresh water. Several genera disappeared in the Oligocene. The Miocene marked the highest development of diatoms with the appearance of many new genera, the Centrales predominating but with the Pennales showing a greater diversity. At present, half of the Miocene genera are still extant and the Centrales have been overtaken by the Pennales.

On the whole, it is a group that has evolved slowly. It is estimated that 15% of present-day species already existed in the Eocene and 6% in the late Cretaceous.

3. MINOR GROUPS

This is a convenient point to return to the siliceous sponge spicules mentioned in Chapter 6. Sediments containing radiolarians and especially diatoms, in many cases also yield small numbers of siliceous microfossils attributed to various groups.

Silicoflagellates

The skeletons of silicoflagellates (Figs 8.69–75) are characteristically geometric in appearance; they range in size from 20 to 100 μm and are formed from hollow tubes with a finely reticulated surface. They are the remains of unicellular chrysophyte algae (order Chrysomonadales, suborder Silicoflagellinae) containing numerous small brown chloroplasts and equipped with both pseudopodia and flagellae. All silicoflagellates are marine, planktonic and can tolerate great variations of salinity (optimum: 30 to 40$^0/_{00}$, minimum 20$^0/_{00}$) and temperature. The thermal optima for extant species are, however, quite distinct: 0°C for *Distephanus octonaria* (Ehrenberg, 1837) and 18 to 20°C for *Dictyocha octonaria* Ehrenberg 1837, and their relative proportions are connected to the temperature of the water.

At its most complex, the skeleton is composed of a basal ring that is

SILICEOUS MICROFOSSILS

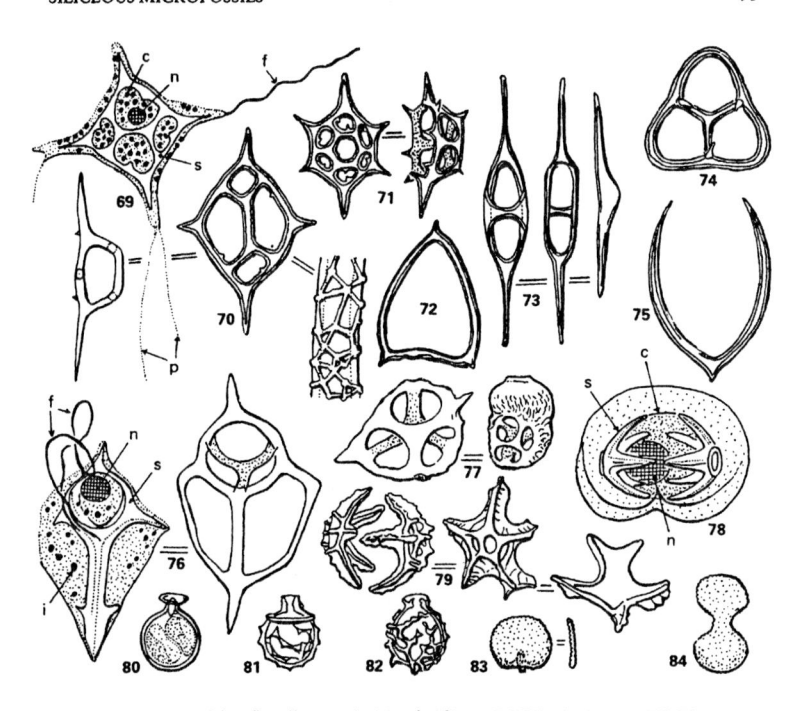

Figs 8.69–84 Silicoflagellates (69–75), ebridians (76,77), *Actiniscus* (78,79), chrysomonads (80–82) and phytoliths (83,84)

69,70 *Dictyocha* (Cretaceous–Recent): 69, living individual; f = flagellum; n = nucleus; p = pseudopodia; s = skeleton (×330); 70, from left to right, skeleton in profile and from above (×330) and detail of reticulated surface (×1200)

71 *Distephanus* (Cretaceous–Recent): from left to right, view from above and profile (×330)

72 *Mesocena* (Eocene–Pleistocene) (×420)

73 *Naviculopsis* (Eocene–Miocene): from left to right, views from above, below and profile (×330)

74 *Corbisema* (Cretaceous–Miocene) (×650)

75 *Lyramula* (Cretaceous) (×400)

76 *Hermesinum* (Eocene–Recent): from left to right, living individual and skeleton; f = flagellum; i = lipidic inclusion; n = nucleus; s = skeleton (×800)

77 *Ebriopsis* (Eocene–Miocene) (×400); on the right, skeleton with cyst (?) (×270)

78,79 *Actiniscus* (Cretaceous–Recent): 78, living gymnodinial Dinoflagellates; c = central capsule; n = nucleus; s = skeleton (×540); 79, different views of the skeleton (×1000)

80,82 Chrysomonadales: 80, living individual in a siliceous cyst (×800); **81,** *Archaeomonas* (Cretaceous–Recent) (×800); **82,** *Outesia* (Eocene–Recent) (×650)

83,84 Two examples of phytoliths: from left to right, (×7000) and (×2000)

After Deflandre, Glezer, Hovasse, Marshall, Perch-Nielsen, etc.

polygonal or elliptical. There is a high degree of intraspecific variation.

The oldest silicoflagellates date from the early Cretaceous (Albian). They become common from the late Cretaceous and are particularly numerous in the Miocene. In total, some 50 species belonging to 15 genera have been described. Only a few species, however, survive at present and the group seems to be on the way to extinction.

Some Rarer Microfossils

The skeletons of the unicellular **ebridians**, measured in tens of micrometres, are very similar to those of radiolarians. They are arranged around a triangular spicule or tetraxon, joined at the extremities to form a closed unit. The skeleton is intracytoplasmic in living organisms which lacks chloroplasts but has a nucleus and two unequal flagellae, which could relate them to the Dinophyceae. Ebridian fossils appear at the end of Cretaceous. The two extant species are planktonic and are found in cold or temperate seas.

The Dinophyceae (Pyrrhophyta) are represented by star-shaped siliceous plates called *Actiniscus* (Figs 8.78 and 79). They were known in fossil form before being discovered grouped in pairs around the cell nucleus of a living species of marine dinoflagellate in the order Gymnodiniales. (See Chapter 10 for a more detailed treatment of dinoflagellates.)

Diatomaceous sediments, whether of marine or lacustrine origin, sometimes yield tiny siliceous shells of from 2 to 30 μm in diameter, spherical or elliptical in shape, and with a pore at the summit of a short neck. They are the cysts of unicellular algae with yellow photosynthetic pigments. Known as **chrysomonads** (Chrysophyta: Chrysomonadales) (Figs 8.80–82), they protect the living cell, the flagellum emerging from the pore. They are equally important as cysts, (usually called **archaeomonads**) known from the late Cretaceous. Their classification is based, for the 150 fossil species, on the general form, ornamentation and appearance of the pore. Living chrysomonads are found in every type of aquatic habitat.

Finally, lacustrine diatomites, palaeosoils and various carbonaceous sediments (peats) contain varying amounts of siliceous particles ranging in size from 1 to 100 μm and taking the form of dumbells, fans, rods or spines with a wall that may be smooth or granular. They correspond to deposits of silica that are either intracellular or situated in or between the cell walls of higher plants (e.g. Equisetaceae and Gramineae). They are termed phytoliths or phytolitharia (Figs 8.83 and 84).

CONCLUSION

Although neglected for many years by palaeontologists, siliceous microfossils were collected during the nineteenth century because of their elegance as works of art. Their geological significance became apparent once seafloor drilling opened up the possibility of examining continuous series of Cenozoic sediments containing siliceous microfossils either alone or in association with coccoliths and planktonic foraminifera.

The systematic study of these microfossils is in the midst of a great expansion and it is difficult to foresee what results will emerge from the research that is under way. It can, nevertheless, be said that radiolarians will be the markers for stratigraphy while diatoms will provide palaeogeographic information on the various lacustrine and marine sediments of the Cenozoic and the Pleistocene. Their stratigraphic range is shown in Fig. 8.85.

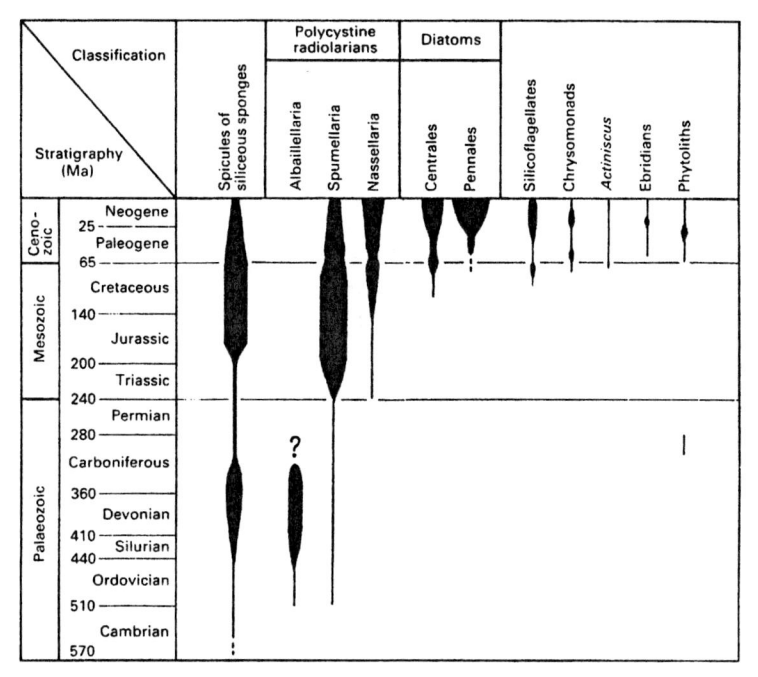

Fig. 8.85 Stratigraphic range of siliceous microfossils

BIBLIOGRAPHY

As there are no recent works covering all radiolarians, the following may be consulted though it uses the old classification of Haeckel: A.A. Strelkov & R.K. Lipman in Y.A. Orlov (ed.): [*Fundamentals of Paleontology: Protozoa*], pp. 552–712 (Academy of Sciences, Moscow, 1959; English translation, Israel Program for Scientific Translations, Jerusalem, 1962). Well illustrated monographs including the following publications: W.R. Riedel & A. Sanfilippo 'Cenozoic Radiolaria', in A.T.S. Ramsay (ed.), *Oceanic Micropalaeontology*, vol. 2, pp. 847–912 (Academic Press, New York, 1977) and E.A. Pessagno Jr, 'Radiolarian zonation and stratigraphy in the Upper Cretaceous portion of the Great Valley Sequence, California Coast Ranges'. (*Micropaleontology Special Publication* 2, 1976). To obtain some idea of older associations, the publications available are more or less restricted to preliminary publications such as: H.P. Foreman *Micropaleontology*, vol. 9, No. 3 (1963) and those of P. De Wever, *Micropaleontology*, vol. 25, No. 1 (1979) and *Revue de Micropaléontologie*, vol. 24, Nos 1, 3 and 4 (1981–1982). H.P. Foreman & W.R. Riedel, *Catalogue of Polycystine Radiolaria* (American Museum of National History Special Publication, New York, 1972–), in course of publication, is intended for specialists. An excellent, authoritative and comprehensive review of knowledge (especially of the fine structure, cytology, physiology and ecology) of radiolarians has been recently provided by O.R. Anderson (*Radiolaria*, Springer-Verlog, 1983).

The literature devoted to diatoms is even more scattered. See, for example, W.W. Wornardt Jr, 'Diatoms, past, present, future', in *Proceedings of the 1st International Planktonic Conference on Microfossils (Geneva, 1967)*, vol. 2, pp. 690–714 (1969), and the chapter devoted to them in H. Tappan, *The Paleobiology of Plant Protists*, pp. 567–677 (W.H. Freeman, San Francisco, 1980). Other works, which though old may be of service to a beginner, are: H. Van Heurck, *Traité des Diatomées* (Anvers, 1899; English reprint, 1962); P. Lefébure, *Atlas pour la Détermination des Diatomées* (Paris, 1947). A more detailed work is S.L. Van Landingham, *Catalogue of the Fossil and Recent Genera and Species of Diatoms and their Synonyms*, 8 vols (Vaduz Cramer, 1967–1979). Recent research on diatoms is presented in Symposiums on *Recent and Fossil Marine Diatoms*, published in *Nova Hedwigia* (Weinheim, FRG).

The silicoflagellates are treated in: Z.J. Glezer, 'Silicoflagellatophyceae', in M.M. Gollerbakh [*Cryptogamic Plants of the U.S.S.R.*), vol. 7 (Academy of Sciences, Moscow, 1966; English translation, Israel Program for Scientific Translations, Jerusalem, 1970); E. Martini 'Systematics, Distribution and Stratigraphical Application of Silicoflagellates', in A.T.S. Ramsay (ed.), *Oceanic Micropalaeontology*, vol. 2, pp. 1327–1342 (Academic Press, New York, 1977) and also in H. Tappan, *op cit.*, pp. 535–566.

Chapter 9

Conodonts

Conodonts, like vertebrate bones, consist of calcium phosphate and this unusual chemical composition distinguishes them from all other microfossils. Given the fact that they occur in comparatively low densities in compact rocks of the Palaeozoic and Triassic, their preparation requires the chemical treatment discussed in Chapter 2.

1. GENERAL ORGANIZATION

They are small structures (Figs 9.1 and 2), which are either opaque or translucent and black or amber in colour. Their length ranges from 0.2 to 6 mm but averages around 1 mm. The density (2.8 to 3.1) is high. Some 83% consists of calcium phosphate crystallized as francolite within a matrix of organic materials. Francolite is a carbonate fluorapatite with the formula:

$Ca_5 Na_{0.14} (PO_4)_{3.01} (CO_3)_{0.16} F_{0.73} (H_2O)_{0.85}$

Depending on their morphology, conodonts fall into three groups (see Figs 9.3–21):
- **Simple** forms (**cones**) with the appearance of a tooth or hook (e.g. *Drepanodus, Acanthodus*).
- **Composite** forms with a principal tooth flanked laterally by denticulate blades (e.g. *Hibbardella* and *Coleodus*).
- **Platform conodonts** which consist of a blade partially bordered by two lateral extensions covered with nodules and ridges. The platform is not very distinctive in *Icriodus* and *Amorphognathus* but exhibits considerable development in *Gnathodus, Ancyrodella, Palmatolepis*, etc.

Conodonts are found in enantiomorphic pairs. On the lower side of each specimen, there is a **basal cavity** which may be covered by a **basal plate**. In section (Fig. 9.2), conodonts and basal plates are seen to have a lamellar

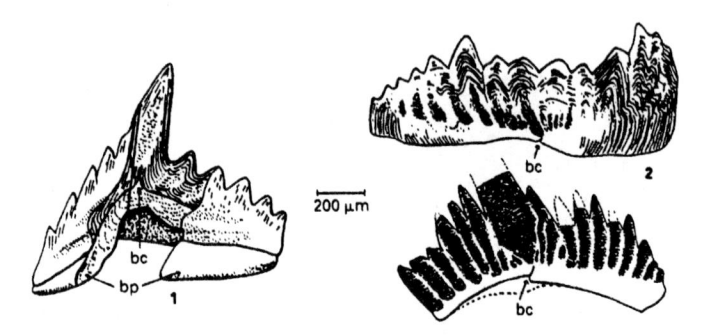

Fig. 9.1 (above) Diagram of a composite conodont with lower side and basal cavity (bc) masked by the basal plate (bp). After Lindstrom (1973, fig. 1)

Fig. 9.2 Internal structure of conodonts: above, *Spathognathodus* (Ordovician–Permian) lamellar structure barely masked by white matter (×35); below, *Ozarkodina* (Silurian–Devonian) completely filled with white matter (×100). After Lindström (1973, figs 4F and 5A)

Figs 9.3–21 (opposite) Primitive conodonts (3, 21), simple conodonts (4, 5), composite conodonts (6–10), primitive platform conodonts (11–13) and true platform conodonts (14–20)

3 *Westergaardodina* (Cambrian–Ordovician) (×30)
4 *Drepanodus* (Ordovician): from left to right, anterior view, section and lateral view (×25)
5 *Acanthodus* (Ordovician) (×25)
6 *Chirognathus* (Ordovician) (×25)
7 *Ligonodina* (Ordovician–Triassic) (×25)
8 *Ozarkodina* (Silurian–Devonian) (×25)
9 *Hibbardella* (Ordovician–Triassic): from left to right: anterior, lateral and posterior views (×12)
10 *Coleodus* (Ordovician) (×25)
11 *Prioniodus* (Ordovician): from left to right, lateral view and view from above (×35)
12 *Icriodus* (Devonian–Carboniferous): from left to right, section, view from below and view from above (×25)
13 *Amorphognathus* (Ordovician–Silurian): from left to right, view from above (×25), lateral view (×50) and view from below (×50) (three different species)
14 *Gnathodus* (Devonian–Permian): from left to right, views from below and above (×15) and lateral oblique view (×30) (two different species)
15 *Polygnathus* (Devonian–Carboniferous): from left to right, views from below and above (×35), sections (×50) and lateral view (×35)
16 *Cavusgnathus* (Carboniferous): from left to right, lateral view and view from above (×40)
17 *Ancyrognathus* (Devonian): from left to right, views from below and above (×25)
18 *Ancyrodella* (Devonian): view from above (×25)
19 *Palmatolepis* (Devonian): from left to right, view from above, from below without basal plate, section and view from below with basal plate (×25)
20 *Gondolella* (Carboniferous–Permian): from left to right, lateral view, and views from above and below (×25)
21 *Pygodus* (Ordovician) (×25)

After Bischoff, Branson, Hass, Lindström, Mehl, Ziegler, etc.

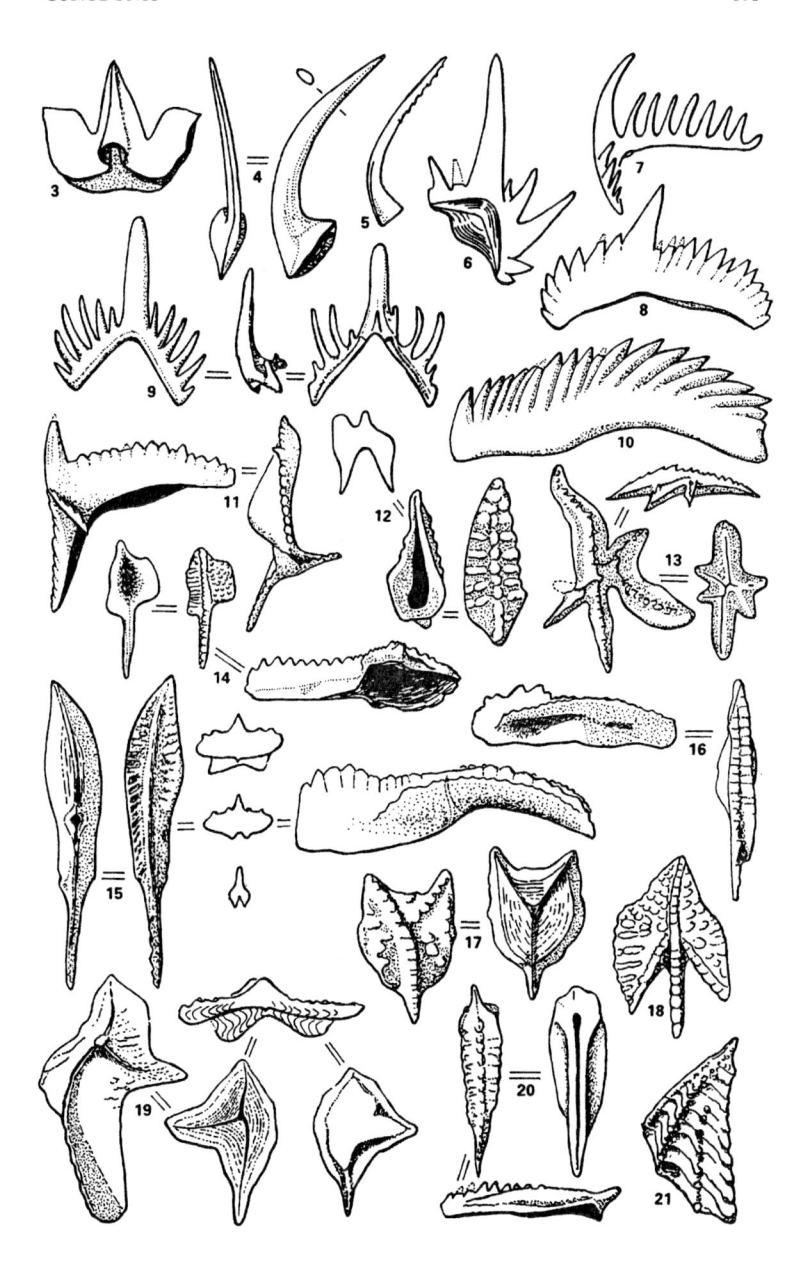

CONODONTS

structure, which, in the case of the former, begins from the basal cavity. Each element consists of around 100 **lamellae**, ranging from 2 to 3 μm in thickness and formed from fusiform crystals with a diameter of 0.5 μm. These lie parallel to each other and perpendicular to the lamella. Growth is centrifugal and involves the accretion of lamellae one above the other, a fact which suggests that the living conodont was surrounded by soft tissue. This deduction tallies with the absence of wear in conodonts and the observation of specimens in which growth has resumed after fracture.

In places, the lamellar structure may be masked by tiny canals or vesicles filled with opaque 'white matter'. Where this occurs, the conodont appears white or grey in reflected light and black in transmitted light. These cavities may result from the reabsorption of lamellae during the life of the organism and the recrystallization of the phosphate into very fine globular crystals.

The classification of conodonts follows a purely morphological parataxonomy. 300 'genera' and several thousands of 'species' have been described. Slight differences have been interpreted as the effect of intraspecific polymorphism.

2. SYSTEMATIC AFFINITIES AND BIOLOGICAL SIGNIFICANCE

Conodont Deposits

Conodonts are found alongside radiolarians, graptolites and fragments of fish in near shore marine facies (sands and glauconitic sandstones) and in deeper facies (fine, more or less dolomitic limestones, black shales, phosphorites and cherts). They are absent or uncommon in reef facies and in facies containing fusulinid foraminifera, crinoids, brachiopods and calcareous algae. As a general rule, the lower the rate of sedimentation, the greater the abundance of conodonts in a rock. On average, a dozen individuals are to be found in a kilogram of rock but, in exceptional cases, there may be as many as several thousand.

The palaeobiogeographical distribution of conodonts is wide. They are found in western Europe, around the Mediterranean, in the Sahara, in India, China, Australia and in North America. Certain genera (e.g. *Icriodus* and *Cavusgnathus*) are found on their own in littoral facies but are associated with other genera (e.g. *Palmatolepis*, *Amorphognathus*, *Ancyrodella* and *Ancyrognathus*) in deep facies. This phenomenon is explained as the result of the stratification of planktonic or nectonic organisms, one group living in shallow water and the other in deeper water (Fig. 9.22).

Assemblages of Conodonts

Current methods of preparation make it possible to obtain isolated conodonts. Nevertheless, from *in situ* examination in sediments, particularly on the stratification planes of schists, it can be seen that they are grouped in symmetrical **assemblages** (Fig. 9.23). The first such discoveries were made

CONODONTS

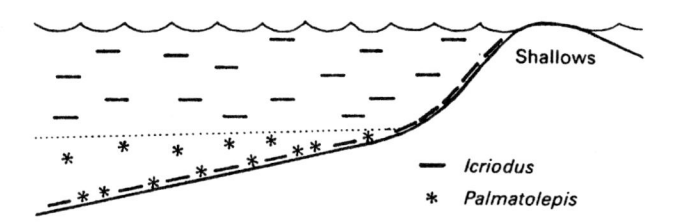

Fig. 9.22 Reconstruction of the stratification of conodontophorid populations in the late Devonian. After Seddon & Sweet (1971, text fig. 1)

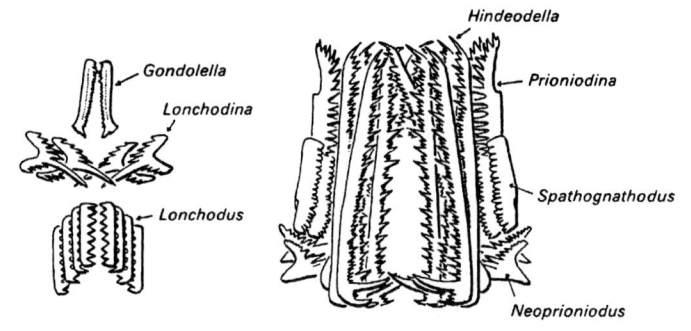

Fig. 9.23 Natural assemblages of conodonts: left, *Illinella* (Carboniferous); right, *Lochriea* (Carboniferous) (×20) After Rhodes (1954).

simultaneously in the USA and Germany. At first it was thought that the groups were chance occurrences but, as subsequent discoveries multiplied in different deposits, it became possible to conclude that the assemblages were natural. Several hundreds are known at the present time. Each one of these contains seven or eight pairs of conodonts that, in isolation, are recognized as belonging to different genera.

Conodontophorids

Although the conodonts' hosts have been designated as a separate group or class – Conodontophorida – the problem of their systematic affinities remains unsolved. Some specialists view conodonts as endoskeletal elements of primitive vertebrates. Given that vertebrates are virtually alone in building an apatite skeleton, this hypothesis had to be taken seriously. Nevertheless, this position should not be accepted unreservedly, since the conodonts cannot be teeth, as their growth is centrifugal and pulp cavities are absent, or bones (dermic plates, elements of a branchial skeleton), as there is no trace of osteoblasts or blood vessels. Finally, their stratigraphic distribution fails to coincide with any vertebrate fossil group.

The recent discovery, in fine laminated Carboniferous limestones in Montana, of conodonts within fossil prints whose appearance and size (6 to 7 cm) suggest those of primitive chordates such as *Amphioxus*, has not helped to clarify the problem. Some palaeontologists feel that the prints correspond to possessors of conodonts. Others, given the disorganized arrangement of the conodonts, explain the prints as being left by species preying on Conodontophorida.

A still more recent discovery is of a print 4cm long in the Scottish Carboniferous which has at its front end a double bulge behind which is located an assemblage of conodonts. Traces of musculature in the form of chevrons, and of an asymmetric caudal fin have led the attribution of this conodontophorid to a particular group of vertebrates, as yet unknown. Other specialists believe it could be a Chaetognatha.

It must be concluded, therefore, that the conodonts remain highly enigmatic. They belong to pelagic marine organisms that deduction shows to have had a soft bilaterally symmetrical body with a length of some centimetres.

The role of the conodonts is equally uncertain. Some authors believe that the assemblages must have been endoskeletal supports for a peribuccal organ comparable to the lophophore of brachiopods, bryozoans and pterobranchs, serving as a sieve and for the concentration of food particles.

3. CONODONTS THROUGHOUT GEOLOGICAL TIME

Conodonts seem to have appeared at the end of the Precambrian (in Siberia) but to have remained few in number during the Cambrian. Most of these, moreover, are **paraconodonts** (*Westergaardodina*) with a particular simple form and they are only weakly mineralized.

The associations found from the base of the Ordovician are very diversified, with 85 to 100 genera being known. Both simple and composite forms abound but platform types are only just appearing. In the Silurian, conodonts are more rare and simple forms are scarcely found after this period. The group reaches its highest development in the Devonian with the spread of platform conodonts. A number of genera (e.g. *Ancyrognathus*, *Ancyrodella* and *Palmatolepis*) are restricted to this period. In the early Carboniferous, the associations are still rich. Decline sets in with the late Carboniferous; by the Permo-Triassic conodonts are rare. New genera cease to appear, the size of individuals is reduced and the basal cavity is missing or relatively unmarked. By the end of the Triassic, conodonts have disappeared from the fossil record.

CONCLUSION

Because of the very rapid succession of genera and species, conodont microfossils are of great stratigraphic value, although their usefulness is limited to the dating of Palaeozoic and Triassic strata. Furthermore, although the Conodontophorida have died out, they have a special place in animal evolution and are of major phyletic interest, if they were indeed primitive vertebrates.

BIBLIOGRAPHY

Although dated, two well-illustrated works deal with the group as a whole: W.H.H. Hass, F.H.T. Rhodes & K.J. Müller, 'Conodonts' in R.C. Moore, *Treatise on Invertebrate Paleontology. Part W: Miscellanea*, pp. W3–W97 (Geological Society of America and University of Kansas Press, Laurence, 1962) and M. Lindström, *Conodonts* (Elsevier, Amsterdam, 1964). The most recent article concerning systematics is by D. Briggs, E. Clarkson and R. Aldridge, The conodont animal, in *Lethaia* 16 (1)1–14 (1983).

Several collective publications treat the stratigraphy and palaeobiology of the conodonts: W.C. Sweet & S.M. Bergström (eds), *Symposium on Conodont Biostratigraphy* (*Geological Society of America, Memoir* no.127, 1971) and F.H.T. Rhodes, *Conodont Paleozoology* (*Geological Society of America, Special Papers* no.141, 1973). It is also worth looking at two recent papers by S.M. Bergström & D.L. Clark in F.M. Swain (ed.), *Stratigraphic Micropaleontology of Atlantic Basin and Borderlands*, pp. 85–136 (Elsevier, Amsterdam, 1977).

Finally, genera and species have been gathered together in a catalogue that is in the course of publication: W. Ziegler, *Catalogue of Conodonts* (Schweizerblart'sche, Stuttgart, 1973– ; 4 vols have appeared).

Chapter 10
Palynology

1. PALYNOLOGY AND PALYNOMORPHS

Palynology – the study of spores and pollen grains – is an interdisciplinary field involving medicine, botany, agronomy, archaeology, and various other subjects. In micropalaeontology, however, the term is used in a more restricted way to designate the study of non-mineralized 'organic' microfossils called **palynomorphs** or **sporomorphs**. All microfossils are of course, of organic origin and when palynologists describe certain varieties as 'organic', they are referring not to the origin but to the chemical composition.

Although non-mineralized, the highly polymerized organic constituents of palynomorphs are extremely durable and are virtually indestructible except by oxidation. Their composition is still poorly understood. It is possible to distinguish:

- **Sporopollenins**, terpenic in nature, the units of which are oxidized carotenoids with the general formula $C_{90} H_{115-158} O_{10-44}$.
- **Chitins**, polysaccharide amines with long linear chains containing hundreds of glucosamine units.
- Subsidiary **lignins**, derived from phenols, and **cutins** that are lipidic.

Wherever sediments have not been subjected to diagenetic oxidation or excessive temperature, they are likely to yield palynomorphs. Dark fine-grained sediments are often extremely rich, with 50 000 or 100 000 per gram of rock; in certain coals, the figure may rise to millions. The originally supple microfossils have been crushed, folded and flattened parallel to the stratification. They may be ambered, transparent and capable of deformation, or black, opaque and brittle with every conceivable intermediate stage.

Palynology is, therefore, a special branch of micropalaeontology based on the common chemical nature of certain microfossils. All these microfossils react in the same way to agents used in preparation and are found together in the residue after the elimination of the mineral phase of the sediment. The residue or **palynofacies** consists of:

- Palynomorph microfossils and nanofossils – spores, pollens and other remains of higher plants, remains of microorganisms (e.g. Dinophyceae) and various metazoans.
- Amorphous organic matter deriving from the partial destruction of the constituents of various cells. The study of this material falls within the province of sedimentary geochemistry and is not of concern here.

2. SPORES AND POLLENS

Biological Significance and Taphonomic Consequences

Spores and pollens[1], the reproductive organs of higher plants, are produced by the **diploid sporophyte** following double division and halving of the chromosome complement of the mother-cell. They are grouped in **tetrads** in the **sporangium** or **antheridium**.

After germination, the spore produces an autonomous **haploid gametophyte** (leafed plant in the Bryophyta (mosses and liverworts), prothallus in the Pteridophyta), which produces the sex cells – **oospheres** and **antherozoids**. The spores may be identical (**isospory** of the Bryophyta, Lycopodiales (club mosses, Eufilicales, etc.) or differentiated into male **microspores** and female **megaspores** (**heterospory** of the Selaginellales, Hydrofilicales, etc.). They function as organs of both reproduction and dissemination.

The pollen of the Spermatophyta (seed-bearing plants) is the equivalent of a microspore. On germination, it produces a male gametophyte of reduced size that produces antherozoids. It functions, however, only as a sexual organ, dissemination being carried out by the seed.

Spores and pollens are emitted by plants that live, for the most part, in areas that are continental and subaerial. As a result of their enormous power of dispersion, they are found as fossils not only in different continental sediments but also in a wide variety of marine deposits. After preparation, fossil specimens are sometimes found still grouped in tetrads but, more often, they are isolated.

General Morphology and Germinal Apertures

Spores have bilateral symmetry with a single **germinal aperture** at the proximal pole (towards the interior of the tetrad) and taking the shape of a rectilinear slit (**monolete type**) or three branches (**trilete type**) (Figs 10.1–14). The size of the spores ranges from a few micrometres to 4 mm. In the absence of biological information, micropalaeontologists have agreed to use the term **microspore** for all specimens under 200 μm; any that are larger are called **megaspores** or **macrospores**. In the former case, it is often not possible to tell from the

---

[1] Strictly speaking, the word 'pollen' designates the substance and so should only be used in the singular. For convenience, however, 'pollen' is often written instead of 'pollen grain' and in this sense may be either singular or plural.

morphology alone whether the subject is an isospore, a true microspore, a small macrospore or even a pollen. For this reason, some authors prefer to use the term **miospore**.

Pollens range in size from 2 to 200 μm. The germinal aperture may be absent (**inaperturate pollen**). If there is only one aperture, (gymnosperms, monocotyledons, some primitive dicotyledons), it is situated at the distal pole (towards the exterior of the tetrad). As a general rule, there are three apertures and these are arranged around the equator. When the number is greater, the

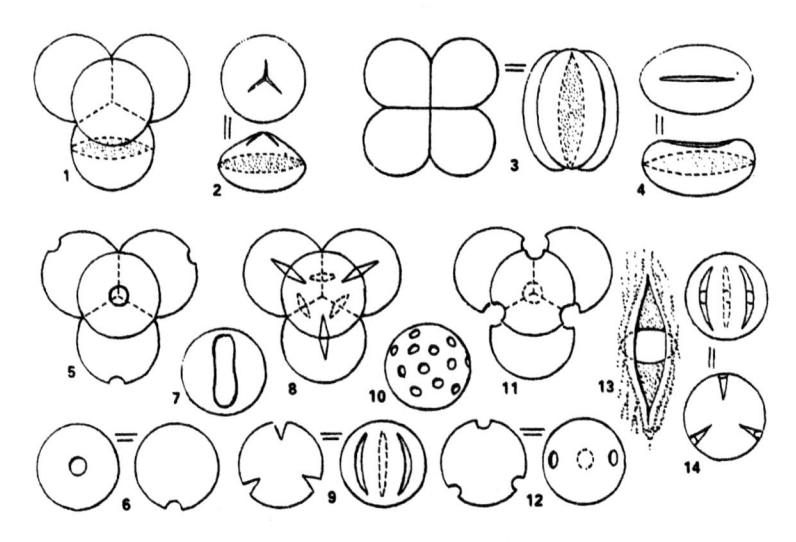

Fig. 10.1–14 Morphology and orientation of spores (1–4) and pollens (5–14): dv = distal view (from the exterior of the tetrad); eqv = equatorial view; pv = proximal view (from the interior of the tetrad)
1 Tetrahedral tetrad of trilete spores (type *Calamospora*)
2 Trilete spore: pv (top) and eqv (bottom)
3 Tetragonal tetrad of monolete spores (type *Laevigatosporites*): view from above (left) and lateral view (right)
4 Monolete spore: pv (top) and eqv (bottom)
5 Tetrad of monoporate pollens
6 Monoporate pollen: dv (left) and eqv (right)
7 Monocolpate (or monosulcate) pollen, dv
8 Tetrad of tricolpate pollens
9 Tricolpate pollen: polar view (= dv, = pv) (left) and eqv (right)
10 Stephanoporate pollen
11 Tetrad of triporate pollens
12 Triporate pollen: polar view (left) and eqv (right)
13 Colporus
14 Tricolporate pollen: eqv (top) and polar view (bottom)

After Erdtman (1963) and Grebe (1971, figs 1 and 2)

apertures are equatorial or are spread irregularly over the surface of the pollen. These apertures are thin areas which, on germination, yield pores or furrows (**colpi**, singular = colpus and **sulci**, singular = sulcus). If the pores and furrows are associated, the pollen is said to be **colporous**.

Structure and Sculpture of the Exine

The wall or **sporoderm** enveloping the cell contents consists of two or three superposed membranes (Figs 10.15-18):

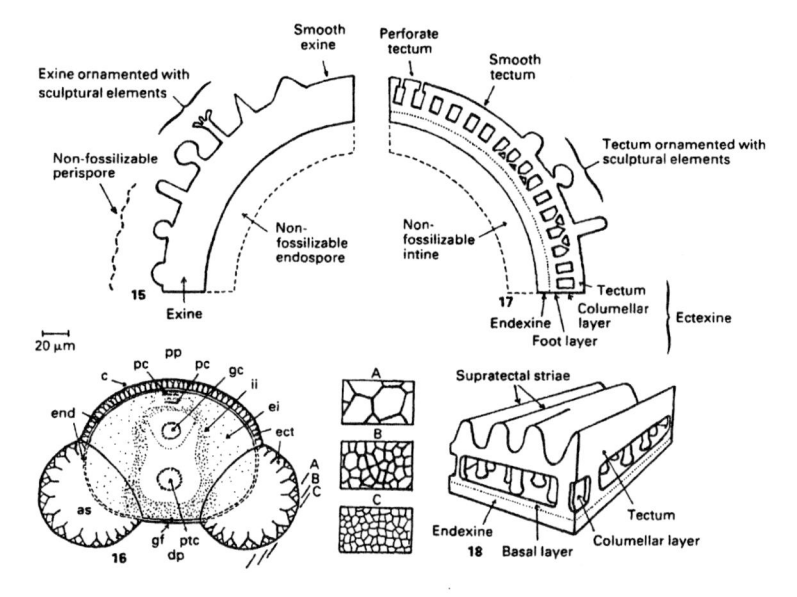

Figs 10.15–18 Structure and sculpture of the sporoderm

15 Spore

16 Diagram of a pine pollen (×295) showing air sac (as); c = cap; dp = distal pole; ect = ectexine; ei = external intine; end = endexine; gc = generative cell; gf = germinative furrow; ii = internal intine; pc = prothallian cells; pp = proximal pole; ptc = pollen tube cell. On the right, views of successive levels in the alveolar structure examined by transparency through the tectum; A, large alveoli (the deepest); B, medium alveoli; C, small alveoli, just below the tectum (×670)

17 Tectate pollen of an angiosperm: the columellae (structural elements) may be simple, or divided at the base or summit

18 Diagram of the sporoderm of a tectate pollen

Diagrams after various authors including Erdtman (1963, pl. 56) and Van Campo & Sivak (1972, pls 10 and 11)

- The superficial **perispore**, particular to spores;
- The **exine**;
- The **endospore** (spores) or **intine** (pollens).

Only the exine, which is impregnated with sporopollenin, undergoes fossilization, the other membranes being pectocellulose in nature. Among present-day spores and pollens, the exine represents between 1.4 and 24% of the total weight of the structure.

The exine may be homogeneous or formed from two layers:

- A superficial layer called the **exoexine** (spores) or **ectexine** (pollens);
- An inner layer called the **intexine** (spores) or **endexine** (pollens).

The external surface of the superficial layer is either smooth or ornamented with various protuberances, described as **sculptural elements**. Many gymnosperm pollens have an ectexine, the internal surface of which is covered with structural elements – **granules** or a network of septulae forming alveoli. This granular or alveolar ectexine is often separated from the smooth endexine by air sacs or vesicles (**saccate pollens**). Tectate pollens, characteristic of most angiosperms, have a complex ectexine with a basal layer and cylindrical or columellar protuberances (structural elements) with extremities that join together to form a **tectum**.

In a given pollen, the peculiarities of sculpture and structure are often different. Distinguishing between them is essential for the correct determination of the pollen and this requires minute observation of ultra-thin sections in the TEM and of optical sections with successive adjustments at different depths in the exine.

Classification and Relations Between Spores, Pollens and Plant Macrofossils

Although every extant species of spermatophyte has a characteristic pollen, it is often difficult to specify the systematic affinities of fossil pollens, particularly when they are pre-Cenozoic. Different methods have been proposed in an attempt to solve this problem:

- Morphological comparison of dispersed fossil spores and pollens with the spores and pollens of living plants.
- Statistical comparison of sporopollenic groups with associations of macrofossils in the same deposit.

Only the discovery of spores and pollens *in situ* with fossil fructifications allows a valid palaeobotanical determination. Despite certain difficulties (e.g. scarcity of suitable material, possibility of synsedimentary contamination and immaturity of spores and pollens), it has been established that, just as with

---

Figs 10.19–37 Trilete (19–26) and monolete spores (27–29), pseudosaccate (30–34) and monosaccate pollens (35–37); dv = distal view; eqv = equatorial view; pv = proximal view
19 *Leiotriletes* (Devonian–Miocene): pv (×250)
20 *Raistrickia* (Devonian–Permian): pv (b = baccules) (×360)
21 *Hystricosporites* (Devonian–Carboniferous): megaspore, pv (b = bifid bristles) (×80)

PALYNOLOGY   111

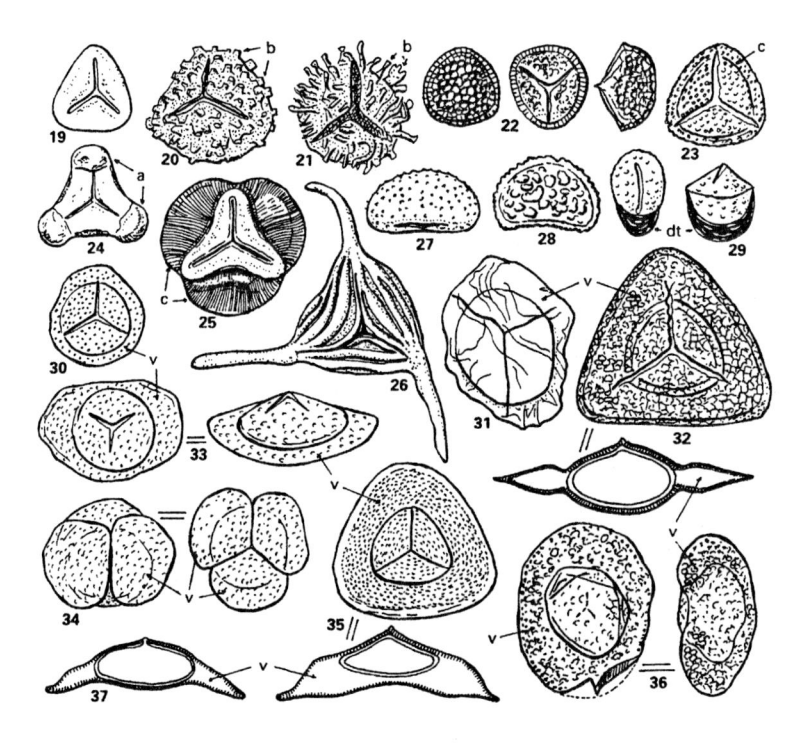

22 *Lycopodium* (Recent): from left to right, dv, pv and eqv (×330)
23 *Lycospora* (Carboniferous–Jurassic): pv (c = cingulum) (×460)
24 *Triquitrites* (Carboniferous–Permian): pv (a = auricles) (×330)
25 *Reinschopora* (Carboniferous–Permian): pv (c = corona) (×400)
26 *Appendicisporites* (Jurassic–Cretaceous): pv showing radial appendices (×330)
27 *Tuberculatisporites* (Carboniferous–Triassic): eqv showing granules (×700)
28 *Polypodium* (Recent): eqv (×270)
29 *Torispora* (Carboniferous): from left to right, oblique pv and eqv (dt = distal thickening) (×380)
30 *Geminospora* (Devonian–Carboniferous): pv (v = vesicle sac) (×230)
31 *Rhabdosporites* (Devonian–Carboniferous): pv (×165)
32 *Spencerisporites* = *Microsporites* (Carboniferous–Permian): megaspore, from top to bottom, pv and axial section (×130)
33 *Schulzospora* (Carboniferous): from right to left, pv and oblique eqv (×280)
34 *Alatisporites* (Carboniferous–Permian): from left to right, eqv and dv (×210)
35 *Endosporites* (Carboniferous–Permian): from top to bottom, pv and axial section (×500)
36 *Florinites* (Carboniferous–Permian): from left to right, pv and eqv (×380)
37 *Nuskoisporites* (Permian–Triassic): axial section (×200)

After Alpern, Doubinger, Erdtman, Kremp, Potonié, Richardson, Schopf, etc.

living plants, there exists, within each of the main fossil groups, a variety of pollen types and the same type may be found in different groups. Certain spores and fossil pollens have thus been attributed, more or less certainly, to plant macrofossils. They have been given the same generic name or, where this is not possible, one that echoes it (e.g. *Ephedripites* (fossil) and *Ephedra* (present)). For most of these microfossils, however, it has been necessary to resort to an artificial morphological classification or parataxonomy. The most widely used is that of R. Potonié and G. Kremp, proposed in 1955 and improved several times since then. Some 600 'genera' are arranged in categories called **anteturmas, turmas, subturmas**, etc.

Spores and Pollens Throughout Geological Time

The earliest known spores date from the Silurian (Llandoverian). They are small smooth forms with trilete markings and derive from Psilophytales. From earliest Devonian times, spores display a significant morphological differentiation (ornamentation, heterospory); at the same time, there appear the pseudosaccate forms, called 'prepollens', attributed either to the pteridophytes (Lycopodiales) or to primitive gymnosperms (Pteridospermales, Cordaitales). They have a central body wrapped in a vesicle formed by the more or less complete separation of the endexine from the ectexine, which has a smooth or granular internal surface.

In the Carboniferous, there is a considerable development and diversification with the successive appearance of several morphological types:

- Monosaccate pollens (Pteridospermales, Cordaitales) derived from the pseudosaccates by the development of an alveolar structure in the exine.

Figs 10.38–63 Pollens – disaccate, striated (38–39), archaic with vestigial proximal scar (40,41), alete (42–46), inaperturate (47,48), polyplicate (49–52), *Classopollis* (53), monocolpates (54–56), Normapolles (57,58), present-day angiosperms (59–63). Same legend as preceding figures with the addition of: v = air-sac; polv = polar view
38 *Striatopodocarpites* (Permian): pv (×470)
39 *Taeniaesporites* (Permian–Lias): pv (×400)
40 *Illinites* (Carboniferous–Permian): pv (×300)
41 *Limitisporites* (Permian): pv (×300)
42 *Phyllocladitites* (Cretaceous–Tertiary): from top to bottom, eqv and dv (×320)
43 *Parvisaccites* (Jurassic–Cretaceous): eqv (×400)
44 *Microstrobus* (Recent) (Podocarpaceae): from left to right, pv, dv and eqv (×480)
45 *Vesicaspora* (Permian): from left to right, eqv and pv (×450)
46 *Abies* (Recent): from left to right, eqv, profile view and dv (×330)
47 *Inaperturopollenites* (Permian–Pliocene) (×750)
48 *Zonalopollenites* (Jurassic–Cretaceous) (×350)
49 *Ephedripites* (Triassic–Eocene): eqv (×400)
50 *Ephedra* (Recent): eqv (×360)
51 *Gnetaceapollenites* (Triassic–Recent): eqv (×400)
52 *Schopfipollenites* (Carboniferous–Permian): from top to bottom, pv and dv (×50)
53 *Classopollis* (Triassic–Cretaceous): from left to right, tetrad (×400), oblique eqv and dv (×650)
54 *Cycadoptites* (Permian–Tertiary): dv (×650)
55 *Ginkgo* (Recent): dv (×650)
56 *Cycas* (Recent): from left to right, dv and eqv (×450)

PALYNOLOGY

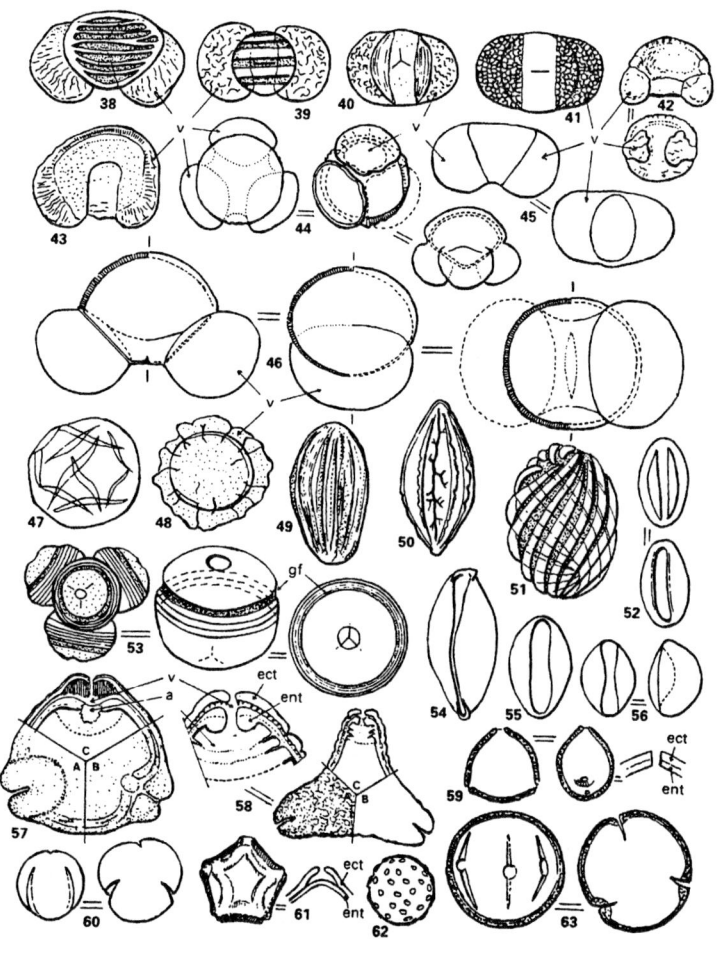

57 *Trudopollis* (Cretaceous–Eocene): polv; A = appearance and external sculpture; B = focus on exine; C = pore structure (×1000)
58 *Atlantopollis* (Cretaceous): same legend; on the left, detail of pore
59 *Corylus* (Oligocene–Recent): from left to right, pv, eqv and detail of pore (×500)
60 *Quercus* (Eocene–Recent): from left to right, oblique eqv and oblique pv (×500)
61 *Alnus* (Oligocene–Recent): from left to right, pv and detail of pore (×500)
62 *Salicornia* (Miocene–Recent) (×500)
63 *Fagus* (Oligocene–Recent): from left to right, eqv and oblique pv (×500)

After Chaloner, Erdtman, Goczan, Hughes, Krutzsch, Lachkar, Potonié

114   PART 1: MICROFOSSIL GROUPS

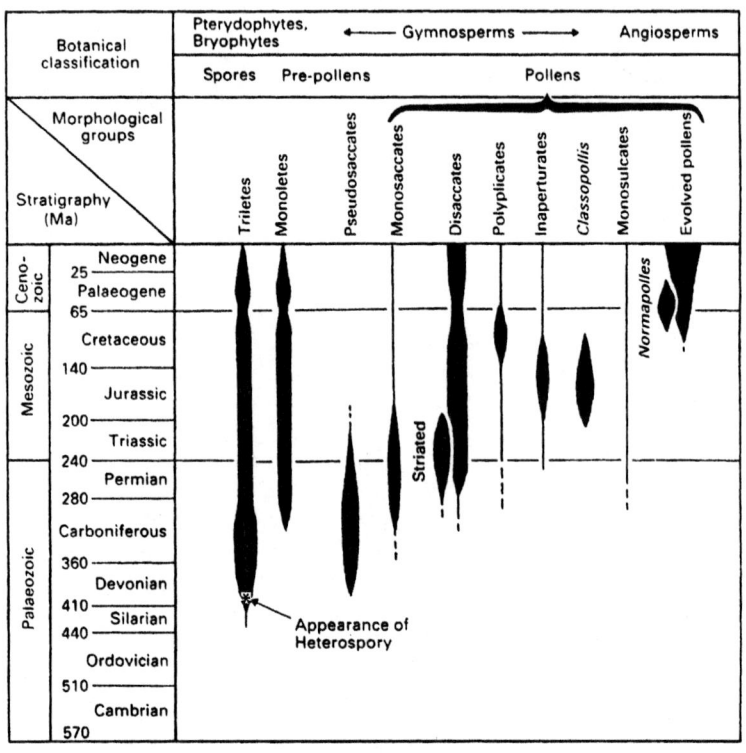

Fig. 10.64 Stratigraphical range of the principal groups of spores and pollens. After various authors

- **Monolete** spores.
- **Disaccate** gymnosperm pollens whose development can be followed from the transformation of certain monosaccates (*Florinites*). They are varied: some are archaic and still bear a trilete or proximal monolete mark; others are striated with rectilinear ridges across the proximal face of the central body (Pteridospermales?); finally, there are the true disaccates, which are close to the pollens of extant conifers.
- The first **polyplicate** and **monosulcate** pollens.

At the boundary between the Carboniferous and the Permian, the spores diminish in importance, giving way to the saccate pollens that dominate in the Permian and Triassic. This was probably linked with a general cooling of the Earth during the Gondwanian glaciations.

A significant renaissance took place at the beginning of the Liassic. Although the pseudosaccates and striated types had disappeared, gymnosperm pollens continued to predominate. They are represented by:

- **Monosaccates** (Cordaitales).
- True **disaccates** (Coniferales).
- **Monosulcates** (Cycadales, Ginkgoales, Bennettitales) with a wide distal furrow.
- **Polyplicates** (gnetophytes – *Gnetum*, *Ephedra* and *Welwitschia*) ornamented with extended longitudinal ridges.
- **Inaperturates** (various Coniferales including Araucariaceae and Podocarpaceae).
- *Classopollis* produced by Coniferales close to Podocarpaceae and possessing both a trilete proximal mark and a distal pseudospore. The representatives of this 'genus' account for up to 99% of the associations found in certain lagoonal levels of the Rhaetian and the Purbeckian.

By Barremian times, the monosulcate pollens have acquired a tetate exine, which makes them resemble the pollens of monocotyledons. The first tricolpatece and tectate pollens belonging with certainty to dicotyledons appear in the Albian or the late Aptian. After the disappearance of *Classopollis* in the Cenomanian, they multiplied greatly. Coinciding with the disappearance of *Classopollis* came the emergence of the triporate Normapolles, a group that persisted into the Oligocene. These pollens are characterized by jutting pores at the level of which there is considerable thickening of the layers of the exine. They have been attributed, though without certainty, to dicotyledons (Amentiflora, Myrtales).

At the end of the Cretaceous, and particularly at the beginning of the Cenozoic, the spores of pteridophytes and the pollens of gymnosperms finally gave way to the angiosperm pollens which, with the exception of the Normapolles, are comparable to those known today.

What took place was, in sum, an attenuation of the pollen through increasing the complexity of the exine. Initially homogeneous, it became alveolar in gymnosperms and then columellar and tectate in angiosperms. Because of the absence or extreme rarity of 'macroremains', several groups of plants (especially the producers of *Classopollis*) are virtually unknown except for their pollens. Thus the study of palynology provides original information to complement and perfect our knowledge of the evolution of the vascular plants.

3. DINOFLAGELLATES

Living Dinoflagellates

Dinoflagellates or Dinophyceae (Pyrrhophyta) are unicellular algae that are morphologically and biologically varied but have in common a very large nucleus (**dicaryon**) and a visible chromosomal apparatus, with yellow or brown **chloroplasts** and two ventral **flagellae** (Fig. 10.65). One of the flagella is axial, smooth and rigid, and is fixed in a short furrow: the **sulcus**. The other flagellum is equatorial, has the appearance of an undulating membrane, and is fixed in a spiral furrow: the **cingulum**.

Sexual reproduction is either absent or poorly understood. Asexual

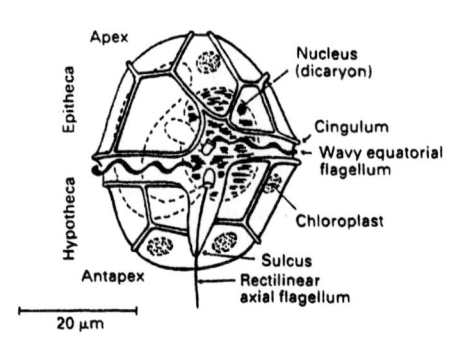

Fig. 10.65 Schematic ventral view of a living Peridinale dinoflagellate with tabulate cellulosic theca (×1050)

reproduction takes place by binary fission.

Certain Dinophyceae lack an external covering (order Gymnodiniales). Others, particularly the order Peridiniales, are protected by a thick coat or theca. This consists of cellulose and is formed from perforate polygonal plates, the number and arrangement (**tabulation**) of which into latitudinal series are constant for every species. Several numbering systems have been proposed. In that of C.A. Kofoid, the complete series (**apical, precingular, postcingular and antapical**) are designated by the indices ', '', ''', ''''. The incomplete intercalary series are given the letters A or P. Within each series, the plates are numbered from 1 to n, starting from the sulcus and moving anticlockwise from an apical viewpoint (Fig. 10.66).

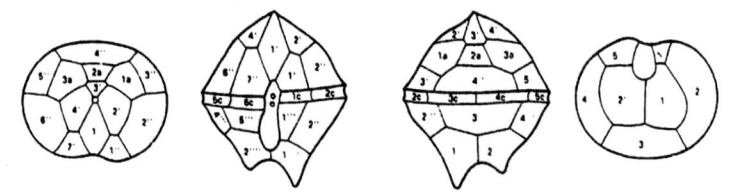

Fig. 10.66 Diagram of the theca of a dinoflagellate (*Peridinium*) illustrating the terminology and the symbols used to identify the different plates. From left to right: apical, ventral, dorsal and antapical views. The tabulation here is expressed by the formula 4'3a7''6c5'''0p2''''. After Evitt (1969, fig. 182)

Many species are parasitic; some are symbiotic (**zooxanthellae**); and others, living independently, are autotrophic plankton of the surface waters of marine, lagoonal and lacustrine environments. In favourable conditions, they may multiply to such a degree that there are several millions of individuals per litre of water, producing 'red tides', in which the waters are full of toxic excretions. Dinoflagellates are a fundamental part of the plankton in the modern environment. In terms of numerical importance, they are second only to diatoms.

Dinocysts and their Fossilization

For a long time, palynomorphs with clear tabulation were considered to be the thecae of the biflagellate cells of Peridiniales. Others, analogous to these thecae but differing in the absence of tabulation and the presence of appendages, were classed as *incertae sedis* in the problematical Hystrichosphaeroideae. Gradually, it was realized that fossil Peridiniales and many hystrichospheres are not the thecae of plant stages but cysts. This view was confirmed by observation (Fig. 10.67) of the cycle of some 50 living species. The cellulosic thecae of plant cells are fragile and do not fossilize. However, cyst walls, formed from lipid constituents that are poorly understood but close to the sporopollenins, are very durable.

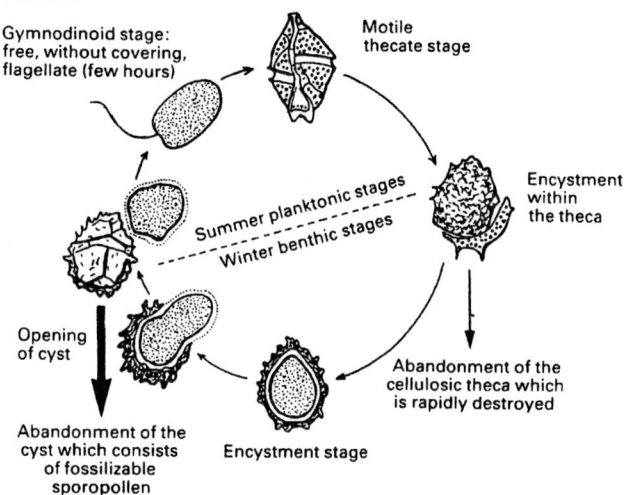

Fig. 10.67 Cycle of *Gonyaulax digitalis* (Pouchet, 1883). The cyst is identical to the fossil species *Spiniferites bentori* Rossignol, 1962. After Wall & Dale (1968, text fig. 2)

Cysts of dinoflagellates – **dinocysts** – range in size from 60 to 150 μm (extremes: 5 and 100). Their systematic classification is based on the spatial relations between the theca and the cyst at the moment the latter is formed (Fig. 10.68).

The **proximate** cyst is emplaced directly against the theca. Their morphologies are very similar but the cyst does not possess true tabulation as it is unable to dissociate into individual plates. Thecal tabulation is indicated (or 'reflected') more or less perfectly at the surface of the cyst by a system or ridges or short spines. The **chorate** cyst is smaller than the theca and is initially joined to it by a system of **appendages** (Fig. 10.69). The way that these processes are arranged on the cyst reflects the thecal tabulation, which can be reconstructed as the appendages originally linked the cyst to the sutures (**sutural tabulation**) or to the centre of the plates (**intrasutural tabulation**) of the theca. The

118　　　　　　　　　　　　　　　　　　　　PART 1: MICROFOSSIL GROUPS

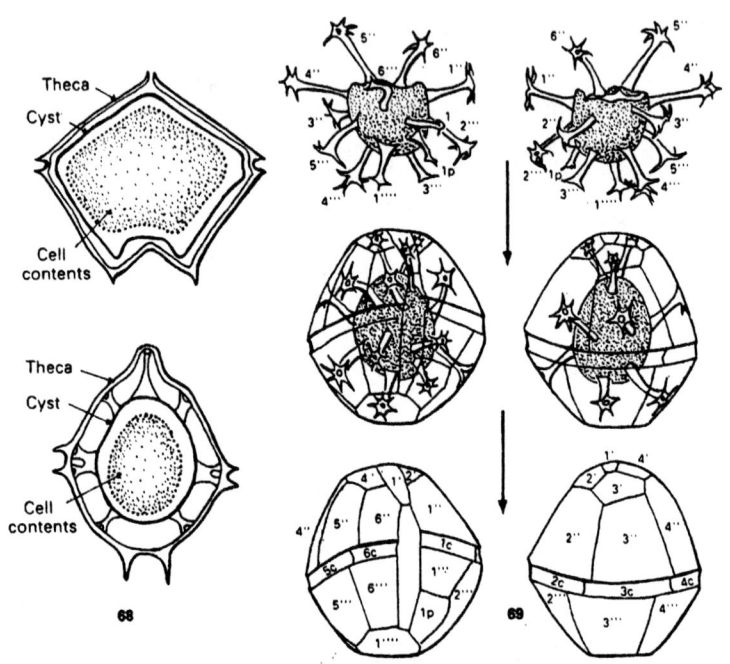

Fig. 10.68 Relative positions of the dinoflagellate cyst and theca: proximate cyst (above); chorate cyst (below). After Evitt (1969, fig. 18.3)

Fig. 10.69 Reconstruction of the form and tabulation (4'6"6c6'''1p1'''') of a theca, similar to that of *Gonyaulax*, from a cyst with intersutural appendices and apical archaeopyle (*Oligosphaeridium*): ventral view (left); dorsal view (right). After Sargeant (1974, fig. 23)

proximate and chorate cysts have a double wall consisting of a thick spongy **periphragm**, lined with a thin dense **endophragm**. A third type, the cavate cyst, has a divided wall, the periphragm forming an outer sphere with a smaller sphere, the endophragm inside it (Fig. 10.70).

A constant characteristic is the position of the **archaeopyle** through which the cell leaves the cyst. Invariably present (Fig. 10.71), this is sometimes a simple fissure situated along a suture but more often has a polygonal contour

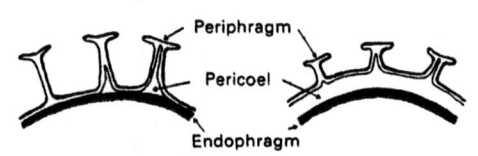

Fig. 10.70 Details of the wall of a chorate cyst (left) and a cavate cyst (right). After Evitt (1969, fig. 18.9)

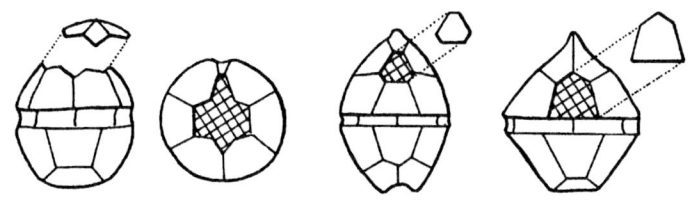

Fig. 10.71 Different types of archaeopyles: from left to right, dorsal and apical views of an apical archaeopyle; intercalary archaeopyle; precingular archaeopyle. After Evitt (1967, figs 16–18)

corresponding to one or several plates. The **operculum** may be detached or may continue to adhere partially to the cyst.

Palynomorphs with clear tabulation and a large number of hystrichospheres with an archaeopyle and a tabulation that can be reconstructed are now categorized as peridinialean cysts. It is hardly surprising that the classification of the 400 'genera' of fossils, established by micropalaeontologists on the basis of dinocyst morphology, far from coincides with the classification of living Peridiniales, arranged by biologists in terms of the anatomy of the biflagellate stages. Cultures of Peridiniales, in fact, show that:

- Species of the same genus develop different cysts.
- A single species may produce cysts that vary slightly in form.
- Several distinct species have identical cysts.
- Only certain species produce cysts. Many others do not form cysts and these include pelagic species which, today, are numerically more significant than any others.

It must be added that fossil dinocysts are known from two other dinoflagellate orders – the Gymnodiniales (*Dinogymnium*) and the Dinophysiales (*Nannoceratopsis*) – although the two groups no longer produce cysts.

Dinocysts Throughout Geological Time

Although dinocysts have been found in the Silurian, the Permian and the Triassic, they were not represented in considerable numbers until the middle Jurassic.

In the Dogger, the predominant form was a proximate cyst with an apical archaeopyle (*Meiourogonyaulax*). This gave way in the late Jurassic to proximate forms with a precingular archaeopyle (*Gonyaulacysta*). They, together with chorate cysts having an apical archaeopyle (*Lithosphaeridium*), are characteristic of the early Cretaceous, a period in which they abounded. The peridinians reach their greatest diversity in the late Cretaceous with great development of genera such as *Spiriferites*, *Palaeohystrichophora*, *Achomosphaera*, *Cordosphaeridium*, *Odontochina* and *Dinogymnium*. The boundary between the Cretaceous and the Tertiary is not particularly marked, numerous Senonian species continuing into the Palaeocene where the predominant forms were *Deflandrea* and *Wetzeliella*. The group has been in slow decline since the Miocene.

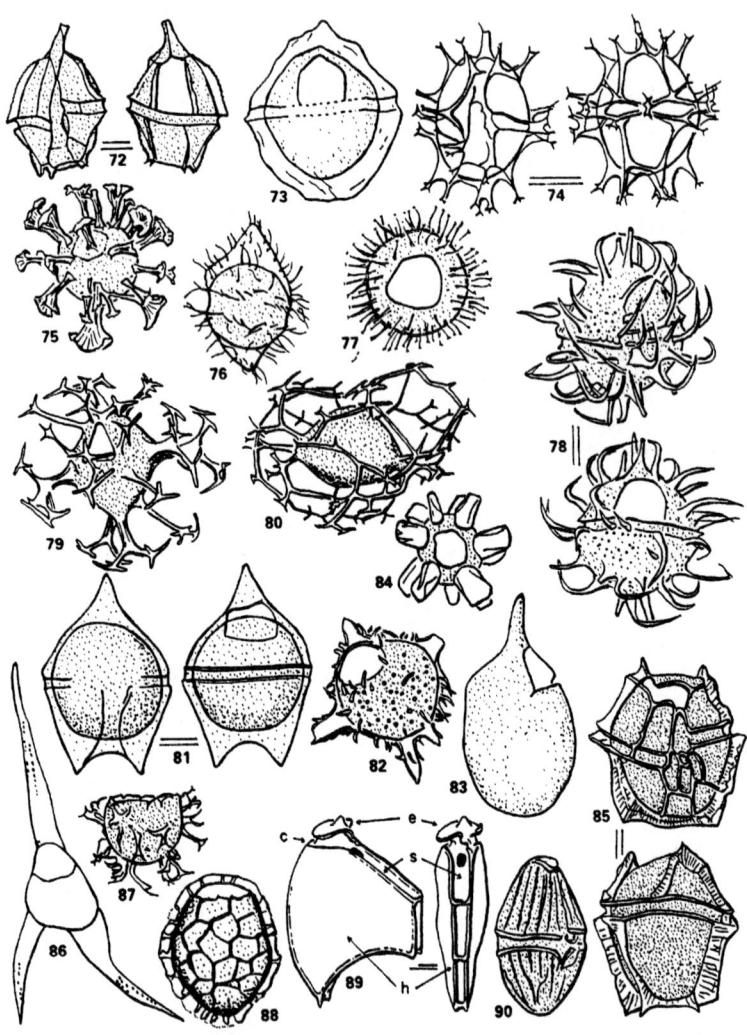

Figs 10.72–90 Peridinialean dinoflagellate cysts with precingular (72–80), intercalary (81–83) or apical (84–88) archaeopyle, dinophysialean (89) and gymniodinialean (90): cav = cavate; ch = chorate; ds = dorsal side; pr = proximate; vs = ventral side
72 *Gonyaulacysta* (Jurassic–Tertiary): pr, vs (left) and ds (right) (×320)
73 *Scriniodinium* (Jurassic–Cretaceous): cav, ds (×270)
74 *Spiniferites* = *Hystrichosphaera* (Jurassic–Recent: ch, vs (left) and ds (right) (×670)

Two lineages have evolved separately since the Lias. The first, the **gonyaulacean** lineage, is based on *Gonyaulax* with a tabulation 3–4'0–3a6''6c6'''1p1'''' (Fig. 10.69). It comprises cysts with a precingular archaeopyle and a tabulation that can almost always be discovered. The second – the **peridinacean** – lineage is based on *Peridinium* with a tabulation 4'2–3a7''3–5c5'''0p2'''' (Fig. 10.66). This covers polyhedral cysts that are ornamented with horns. The surface is often smooth, the tabulation unclear and the archaeopyle intercalary. Cysts with an apical archaeopyle, though common to both lineages, are today found only in the second group. The points of maximum diversification for the two groups are somewhat different. For gonyaulaceans, it lies in the late Jurassic; for peridinaceans, it is rather later.

Many dinocysts derive from marine species. Others belong to lagoonal plankton. Thus, in the Eocene of the Paris Basin, there is a gonyaulacean marine association containing *Spiniferites* and a peridinacean lagoonal association containing *Wetzeliella*. Finally, some dinocysts have been found in Cenozoic lacustrine sediments.

4. ACRITARCHS

The acritarch group was created by W.R. Evitt in 1963 to bring together those hystrichospheres that could not be included with the Dinophyceae.

Acritarchs occur as vesicles ranging in size from 50 to 100 μm, and in exceptional cases from 1 to 500 μm. They are very variable in form, ranging from spherical to cubic. The surface may be smooth, granular or ornamented with extensions of different aspect and arrangement. The central cavity is either closed or communicates with the exterior via a pore, slit or a round aperture, the **pylome**. The chemical composition of the wall is probably identical to that of the dinocysts. It is sometimes simple but is more often formed from two layers either joined together or separated by a cavity or

---

75 *Cordosphaeridium* (Cretaceous–Tertiary): ch, ds (×220)
76 *Paleohystrichophora* (Cretaceous): cav (×570)
77 *Operculodinium* (Tertiary): ch, ds (×270)
78 *Hystrichodinium* (Jurassic–Cretaceous): ch, vs (above) and ds (below) (×500)
79 *Achomosphaera* (Cretaceous–Tertiary) ch, ds (×500)
80 *Cannosphaeropsis* (Jurassic–Tertiary): ch (×380)
81 *Deflandrea* (Cretaceous–Oligocene): cav, vs (left) and ds (right) (×200)
82 *Wetzeliella* (Eocene–Oligocene): cav, ds (×210)
83 *Pareodinia* (Jurassic–Recent): pr, lateral view (×500)
84 *Litosphaeridium* (Cretaceous–Eocene): ch, apical view (×220)
85 *Meiourogonyaulax* (Jurassic): pr, vs (above) and ds (below) (×450)
86 *Odontochitina* (Cretaceous): cav (×200)
87 *Chiropteridium* (Oligocene) (×180)
88 *Valensiella* (Jurassic): pr (×470)
89 *Nannoceratopsis* (Jurassic): lateral (left) and ventral views (right) (×280); c = cingulum; e = epitheca; h = hypotheca; s = sulcus

After Davey, Deflandre, Evitt, Gocht, Sarjeant, Williams, etc.

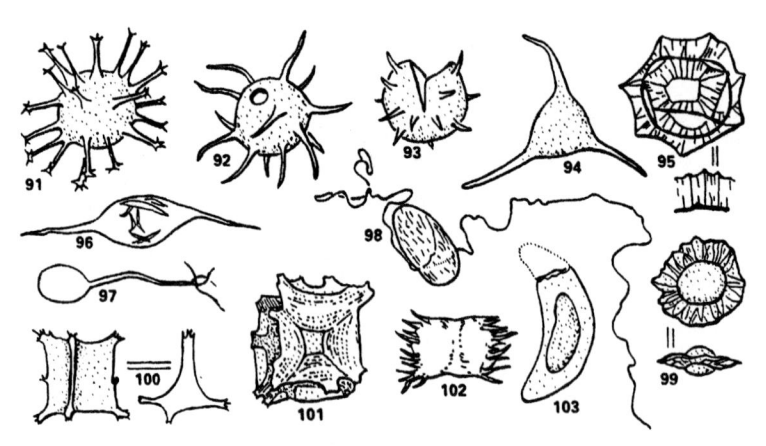

Figs 10.91–103 Morphology of some acritarchs.
91, 92 *Baltisphaeridium* (Cambrian–Recent): 90 (×380); 91 with pylome (×200)
93 *Micrhystridium* (Cambrian–Recent): dehiscence slit (×450)
94 *Veryachium* (Cambrian–Recent) (×400)
95 *Cymatiosphaera* (Ordovician–Tertiary): superficial reticulation (×600)
96 *Leiofusa* (Cambrian–Recent) (×400)
97 *Deunffia* (Ordovician–Silurian) (×170)
98 *Dimastigobulus* (Cretaceous): ophiobolacean with very long appendages (×1000)
99 *Pterospermopsis* (Ordovician–Tertiary): below, profile view (×940)
100, 101 *Polyedryxium* (Ordovician–Devonian): two different species. 100, external view and section (×450); 101 (×380)
102 *Acanthodiacrodium* (Cambrian–Ordovician) (×300)
103 *Wallodinium* (Jurassic–Cretaceous): cavate (×250)

After Kramer, Deunff, Deflandre, Diez, Eisenack, Evitt, Loeblich & Tappan, etc.

pericoel which envelopes the central body.

They are the oldest known fossils. The first, which will be discussed later (Chapter 13), are found in rocks more than 3000 million years old. They are simple spherical forms. Spiny ornamentation appears a little before the Palaeozoic and the pylome in the Upper Cambrian. The group is well represented in the Ordovician and the Silurian but goes into a sharp decline at the beginning of the Carboniferous. Although there was a slight recovery during the Jurassic, the decline has continued since then.

The organisms to which acritarchs belonged remain unknown. The microfossils may be cysts, eggs or remains of unicellular or multicellular animals or plants. A number of features incline specialists to the view that many of the acritarchs are cysts of dinoflagellates, especially Gymnodiniales. Such features include the general form, the nature and composition of the wall and the frequent arrangement of the appendages into polygonal areas or in a pattern reminiscent of tabulation. Given the mystery surrounding their affinities, the classification of the 400 'genera' is inevitably artificial and exclusively dependent on morphology.

The aquatic acritarchs are almost all marine and certainly planktonic. Some Quaternary species lived in fresh water.

The acritarchs form a group not so much because of their size and chemical composition but because of the lack of knowledge of their biological significance. It is a group that, in the words of its creator, stands as a monument to our ignorance and it is one that will disappear when, if ever, the biological relationships are established for all the various organic structures that have been subsumed under it.

5. CHITINOZOANS

Morphology

The Chitinozoa are small sacs or **vesicles** that range in length from 50 μm to 2 mm. Their surface may be smooth, striated, granular, spongy, spiny or ornamented with branching appendages. The test encloses a central cavity or **chamber**, the shape of which may be spherical, ovoid, cylindrical or conical. The opening is via a **pseudostome** situated on the chamber or at the top of a **neck** which flares out to form a **collar**. The flexure joining chamber and neck or chamber and collar is more or less marked. The pseudostome is covered by a disc, the **operculum**, which may be carried by a ringed tube or **prosome**. The base of the vesicle can be convex, pointed or flat. It is sometimes smooth and sometimes surrounded by a crown of appendages. In many cases, it is extended by a tubular **copula** (= mucron, = stolon) the tip of which may dilate to form a peduncular disc (Figs 10.104–115).

The nature of the tegument is uncertain; although chitin-like it does not react in the same way as chitin. It is, however, similar to that of graptolites. When well preserved, it is formed from a non-perforate **ectoderm**, adhering to the outside of which is a thin **periderm** forming the ornamentation (folds, tubular appendages, etc) and which, if removed, may give rise to an **aboral ampulla**. The operculum and the prosome are formed from a different tegumentary membrane, the **endoderm**.

Chitinozoans are generally observed as isolated specimens but most, if not all, were, before fossilization, assembled in linear colonies. The links between individuals were either simple juxtapositions or true joints (Figs 10.110–115). Strong joints, involving the use of the copula, are frequently maintained in preparations.

There are hundreds of species of Chitinozoa distributed among 50 genera.

Palaeoecology and Systematic Affinities

Chitinozoans are found particularly in the fine-grained rocks of anoxic marine facies – shales, sandstones, ferruginous phosphatic and glauconitic sediments. There are at most a few dozen individuals per gram of rock. The original biotopes may have been more varied. Their geographical dispersion (northern and western Europe, Sahara, North and South America) and their association with orthocerid nautiloids, graptolites and acritarchs, all suggest that their mode of life was planktonic or nektonic.

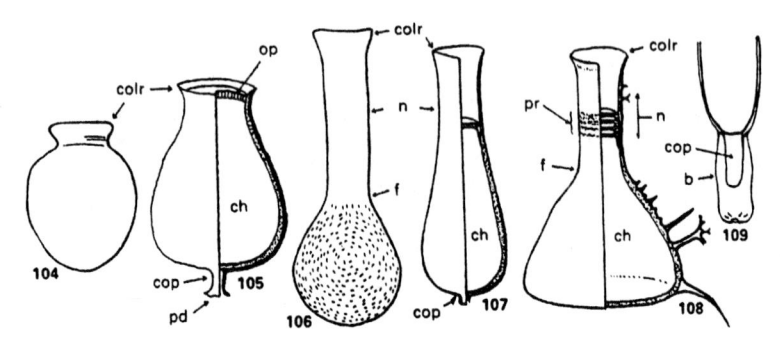

Figs 10.104–109 Morphology and structure of some chitinozoans: b = aboral ampulla; ch = chamber; colr = collar; cop = copula; f = flexure; n = neck; o = operculum; pd = peduncular disc; pr = prosome. Periderm in black, ectoderm stippled, endoderm (operculum + prosome) hatched
104 *Desmochitina* (Ordovician–Devonian): no neck, collar (×210)
105 *Urnochitina* (Silurian): no neck, collar, copula (= mucron) and peduncular disc (×210)
106 *Sphaerochitina* (Silurian–Devonian): cylindrical neck, spiny surface (×210)
107 *Conochitina* (Ordovician–Devonian): long neck, copula, (= mucron) (×105)
108 *Ancyrochitina* (Ordovician–Devonian): cylindrical neck, bristles and aboral crown of tubular appendices, ringed prosome (×315)
109 *Eremochitina* (Ordovician–Devonian): detail of aboral section with membranous ampulla (formed by the expansion and removal of the periderm) surrounding the copula (= stolon) (×210)

After several authors including Paris (1981, fig. 56)

The biological significance of chitinozoans has given rise to diverse hypotheses. Their colonial habit and especially the fact that each chitinozoan has its contents hermetically sealed against the surrounding environment, both lead to the suggestion that they were envelopes for reproductive (eggs) or resting (cysts) stages, not opening until maturity or until the return of favourable conditions. It has been supposed that chitinozoans were formed by graptolites, annelids, gastropods, etc., but the arguments put forward are not very convincing. Less precise but probably closer to the truth is the assertion that these microfossils were produced by soft-bodied vermiform metazoans with a pelagic mode of life.

Chitinozoans Throughout Geological Time

Appearing at the base of the Ordovician (Tremadoc), they become abundant and diversified by the Arenig and until the end of the Silurian. The main evolutionary trends involve a reduction in size from around 500 μm to 120 to 50 μm, an individualization of the body chamber, an increased complexity of ornamentation and a reinforcement of colonial joints. Many genera became extinct at the end of the Silurian and few appear in the Devonian. The group finally disappear in the Strunian (earliest Carboniferous). Because of their rapid evolution, chitinozoans are an excellent stratigraphic tool.

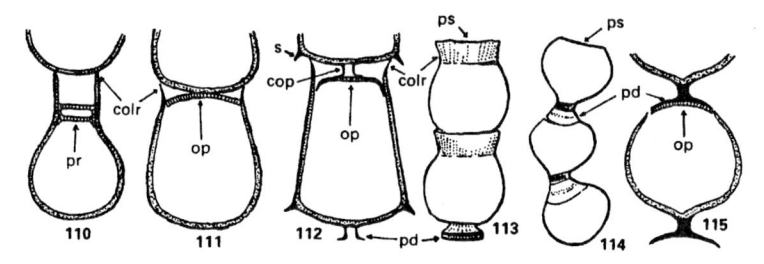

Figs 10.110–115 Colonial arrangements of chitinozoans: colr = collar; cop = copula; op = operculum; pd = peduncular disc; pr = prosome; ps = pseudostome; s = segment. Periderm in black, ectoderm stippled, endoderm (operculum + prosome) hatched
110 Simple juxtaposition, section
111 Double adherence without copula, section
112 Double adherence with copula and segment, section
113 Double adherence with copula and without segment, external view (×50)
114 Reinforced link, external view (×50)
115 Reinforced link, section

After Koslowski (1963, figs 6C and 7B) and Paris (1981, fig. 59).

6. MINOR GROUPS

In addition to the preceding groups, palynofacies contain rarer microfossils. Some of these are without much particular interest (e.g. lignitic fragments with tracheids, cuticular linings of leaves, although suggestive of, e.g. a non-marine environment) but others deserve a more detailed mention.

Remains of Unicellular and Colonial Algae

Being locally very abundant, the **Tasmanitaceae** (e.g. *Tasmanites*) contribute to the formation of bituminous shales (Palaeozoic tasmanites), a source rock for petroleum. They consist of small spheres (30 to 600 μm) flattened by the weight of sediments (Figs 10.116–117). The wall, which consists of a variety of sporopollenin called **tasmanine**, is thick, smooth or slightly mamillate. They are considered to be cysts of marine chlorophyte algae belonging to the Prasinophyceae, because they are unicellular and flagellate, like *Pachysphaera*, which is today found in the plankton of Arctic seas. An identical origin is suggested for the thin-walled ovoid or spherical forms called *Leiosphaeridia*.

Colonial **Chlorococcales** (Chlorophyta) are known from the end of the Precambrian. The accumulation of lacustrine *Pila* gave rise to the algal coals or bogheads (Figs 10.118–119). These algae are identical to present-day *Botryococcus*. Flat colonies, common since the early Cretaceous in lacustrine and marine facies, are attributed to other colonial Chlorococcales, the Hydrodictyaceae, such as *Pediastrum* found today in lakes.

Remains of Protozoa and Metazoa

After palynological treatment, foraminifera are decalcified but their chitinoid basal layers remain in the preparations. These are called 'microforaminifera' and are known from the Lias onwards. Tintinnid loricae are also found although, because of their larger size, they are more easily seen in wash residues (Fig. 10.120).

Metazoans are present in the form of embryo thecae (prosiculae) of graptolites, the (exuvial?) teguments of eurypterid arthropods and scolecodonts (Figs 10.121–123). The last group is used for small chitinoid fragments, measuring from 2.5 to 6 mm, which are recognized as being elements of the masticatory apparatus of polychaete annelids. Generally occurring in isolation, these microfossils are sometimes found in almost symmetrical assemblages reminiscent of the mandibles of eunicid polychaetes. Scolecodonts are fairly abundant from the Ordovician to the Devonian but are subsequently very rare.

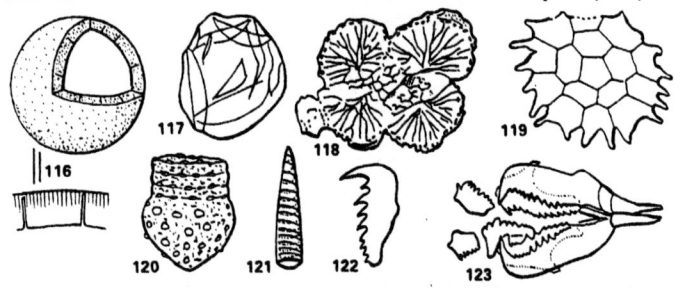

Figs 10.116–123 Some palynomorphs of diverse affinities
116 *Tasmanites* (Precambrian–Recent) (×30); below, section of the wall showing the incomplete pores
117 *Leiosphaeridia* (Precambrian–Permian) (×180)
118 Colony of *Pila* (= *Botryococcus*) in a Permian boghead at Autun (×105)
119 *Pediastrum* (Cretaceous–Recent) (×52)
120 Tintinnid fossil (×335)
121 Graptolite prosicula (×35)
122 Individual scolecodont (×22)
123 Assemblage of scolecodonts (*Paulinites*, Devonian) (×9)

After Bertrand, Bulman, Cramer, Gray, Lange, Kielan-Jaworowska, etc.

Fig. 10.124 Stratigraphic range of some palynomorph groups (not including spores and pollens)

CONCLUSION

Although often mixed in sediments – especially those of marine origin – palynological microfossils derive from two distinct environments:

- Subaerial and lacustrine continental environments – the source of spores and pollens.
- Marine environments – the main source of dinocysts and chitinozoans.

PALYNOLOGY

Because of their wide dispersion, spores and pollens are excellent indicators of continental palaeoenvironments, whereas chitinozoans and dinocysts are good stratigraphical markers. Geologists, therefore, ask a good deal from palynologists who, for their part, have acquired a relatively autonomous position within micropalaeontology. The reason for this is that the objects of their research and their methods of preparation allow them to tackle problems that are outside the normal scope of the micropalaeontologist. Such problems include the evolution of vascular plants and the geological development of organic matter.

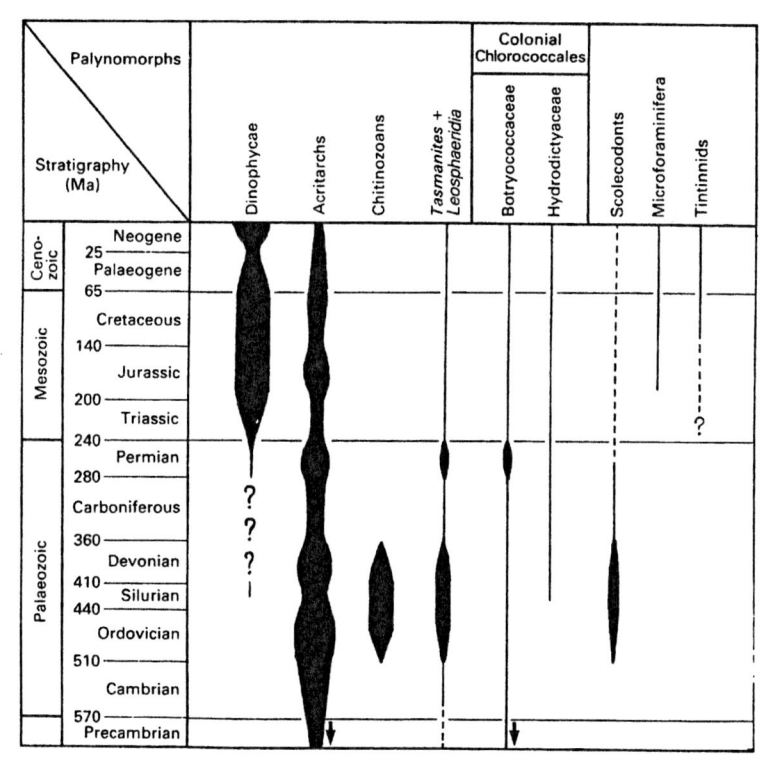

Fig. 10.124 Stratigraphic range of some palynomorph groups (not including spores and pollens)

BIBLIOGRAPHY

All aspects of palynology, including various more or less up-to-date developments, are treated in G. Erdtman, *Handbook of Palynology* (Munksgaard, Copenhagen, 1969), R.M. Tschudy & R.A. Scott (eds) *Aspects of Palynology* (Wiley, New York, 1969) and in J.J. Chateauneuf & Y. Reyre, *Eléments de Palynologie* (Bureau de Recherches Géologiques et Minières (BRGM), Orléans, 1974; typewritten course manual).

Several periodicals are devoted, wholly or in part, to both present-day and fossil palynology: *Grana* (Stockholm, 1954–), *Pollen and Spores* (Paris, 1959–), *Review of Palaeobotany and Palynology* (Amsterdam, 1967–) and *Palinologia* (Leon, Spain, 1979–).

A. Pons, *Le Pollen, Que Sais-je?* no. 783, 2nd edn, (P.U.F., Paris, 1970) is a dense and brief synthesis on spores and pollens. This is complemented by I.M. Pokrovskaia, *Analyse pollinique* (Institute of Mineralogy and Geology, Moscow, 1950; French translation by BRGM 1958) and R. Potonié 'Synopsis der Gattungen der Sporae dispersae', *Beihefte zum Geologischen Jahrbuch, Hanover*, vols 23, 31 and 39 (1956–1960) G.D.W. Kremp, A. Traverse *et al. Catalogue of Fossil Spores and Pollen*, 31 vols (Pennsylvania State University, University Park, 1957–) is for specialists.

For the Dinophyceae, the student should consult W.A.S. Sarjeant, *Fossil and Living Dinoflagellates* (Academic Press, London, 1974). Dinoflagellates and acritarchs are extensively treated in H. Tappan, *The Paleobiology of Plant Protists*, pp. 148–462 W.H. Freeman, San Francisco, 1980). Specialists use A. Eisenack et al., *Katalog der fossilen Dinoflagellaten, Hystrichosphären und verwandten Mikrofossilien* (Schweizerbart'sche, Stuttgart, 1964–; 6 vols and supplements have appeared), and G.J. Wilson and C.D. Clowes, *A Concise Catalogue of Organic-Walled Fossil Dinoflagellates Genera* (New Zealand Geological Survey, Rep. No. 92, 1980).

The most recent works on chitinozoans are those of J. Jansonius, Classification and stratigraphic application of Chitinozoa, in: *Proceedings of the N. American Palaeontological Convention 1969*, (Part G, pp. 789–808, 1970) and F. Paris, 'Les chitinozoaires dans le Paleozoique du Sud-Ouest de l'Europe', *Mémoires de la Société géologique et minéralogique de Bretagne*, Rennes, no. 26 (1981).

Chapter 11

The Position of Microfossils in the Systematic Classification of the Living World

The classification that has been followed in the preceding chapters is an arbitrary one and stems from the main specialisms of micropalaeontology in which microfossils are grouped according to their dimensions and chemical composition. The importance attached to any one group varies greatly and is dependent primarily on its geological and palaeobiological significance (which will form the subject matter of the following chapters). This is not the only factor, however, that affects the presentation. The attention that is given is proportional to the sum of knowledge available. For example, it has not been possible to find a place for bacteria because virtually nothing is known about them even though they have been a subject of interest for many years.

Through the efforts of micropalaeontologists, it has been possible to establish the systematic position of most microfossils (Fig. 11.1). Certain groups, however, remain enigmatic:

- Calpionellids, *Nannoconus*, umbellinids and tentaculitids are associated respectively with the tintinnids, the Coccolithophores, the charophytes and pteropod gastropods, on the basis of vague morphological analogies that have no real meaning.
- Conodonts and chitinozoans derive from organisms that can only be imagined.
- Calcispheres and *Schizosphaerella* could be the remains of animals or plants.
- Acritarchs form a group to which palynomorphs of various origins are provisionally assigned.

Although these uncertainties remain, it can be seen that the diversity of the microfossils is not merely a matter of chemistry, morphology and dimension but also one involving systematic classification.

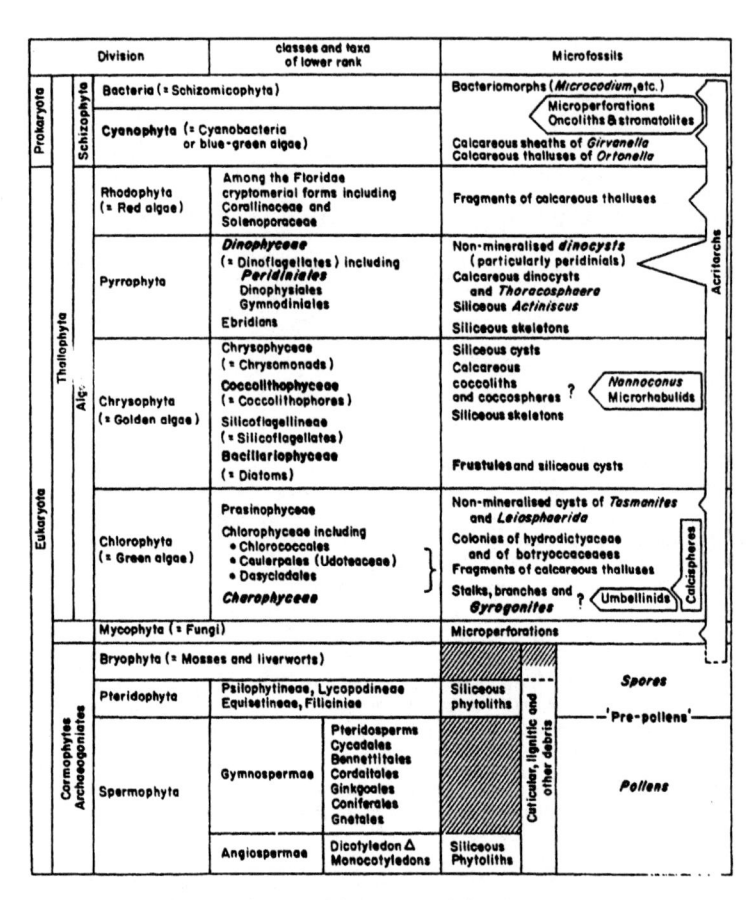

		Division	classes and taxa of lower rank	Microfossils	
Prokaryota	Schizophyta	Bacteria (= Schizomicophyta)		Bacteriomorphs (*Microcodium*, etc.)	
		Cyanophyta (= Cyanobacteria or blue-green algae)		Microperforations, Oncoliths & stromatolites, Calcareous sheaths of *Girvanella*, Calcareous thalluses of *Ortonella*	Acritarchs
Eukaryota	Thallophyta / Algae	Rhodophyta (= Red algae)	Among the Floridae cryptomerial forms including Corallinaceae and Solenoporaceae	Fragments of calcareous thalluses	
		Pyrrophyta	Dinophyceae (= Dinoflagellates) including Peridiniales, Dinophysiales, Gymnodiniales; Ebridians	Non-mineralised *dinocysts* (particularly peridinials); Calcareous dinocysts and *Thoracosphaera*; Siliceous *Actiniscus*; Siliceous skeletons	
		Chrysophyta (= Golden algae)	Chrysophyceae (= Chrysomonads); Coccolithophyceae (= Coccolithophores); Silicoflagellineae (= Silicoflagellates); Bacillariophyceae (= Diatoms)	Siliceous cysts; Calcareous coccoliths and coccospheres ? *Nannoconus* Microrhabulids; Siliceous skeletons; Frustules and siliceous cysts	
		Chlorophyta (= Green algae)	Prasinophyceae; Chlorophyceae including • Chlorococcales • Caulerpales (Udoteaceae) • Dasycladales; Charophyceae	Non-mineralised cysts of *Tasmanites* and *Leiosphaerida*; Colonies of hydrodictyaceae and of botryococcaceaes; Fragments of calcareous thalluses; Stalks, branches and *Gyrogonites* ? Umbellinids	Calcispheres
		Mycophyta (= Fungi)		Microperforations	
	Cormophytes / Archaeogoniates	Bryophyta (= Mosses and liverworts)			Spores
		Pteridophyta	Psilophytineae, Lycopodineae, Equisetineae, Filicinieae	Siliceous phytoliths	
		Spermophyta	Gymnospermae: Pteridosperms, Cycadales, Bennettitales, Cordaitales, Ginkgoales, Coniferales, Gnetales	Cuticular, lignitic and other debris	—'Pre-pollens'— Pollens
			Angiospermae: Dicotyledon △ Monocotyledons	Siliceous Phytoliths	

Fig. 11.1 Simplified classification of the animal and plant kingdoms. It includes only those divisions that are necessary for an understanding of the systematic position of Phanerozoic microfossils. The chitinozoans are too problematic for inclusion in the scheme

MICROFOSSILS IN SYSTEMATIC CLASSIFICATION

Phylum			Classes and taxa of lower rank	Microfossils	
Protozoa	Sarcodina		Rhizopoda including **Foraminiferida**	**Calcareous or agglutinate tests** 'chitinoid basal layers	
			Actinopoda including **Actinopoda**	**Siliceous tests**	
	Ciliata (Ciliophora)		Spirotricha including Tintinnina	Chitinous loricas ?	**Calpionellids** Calpionellomorphs
Metazoa	Invertebrata		Archaeocyathida		
			Porifera (= Sponges)	Calcarea (= Calcisponges) including Stromatoporoidea Hexactinellida (= Silicisponges) Demospongea including Lithistida	Calcareous spicules Siliceous spicules
			Cnidaria (Coelenterata)	Hydrozoa including Hydrocoralliaria Anthozoa • Alcyonaria • Madreporaria	Calcareous sclerites
			Annelida	Polychaeta • Errantia • Sedentaria (Serpulidae)	Coprolites Chitinoid scolecodonts
			Lophophorian	Ectoprocta (= Bryozoa s.s.) Brachiopoda	
			Mollusca	Gastropoda including Pteropods Scaphopoda Bivalvia (Lamellibranchia) including Rudist Cephalopoda	Calcareous shell fragments Coprolites 'Filaments' and calcareous prisms Rhyncolites ? Tentaculitids
		Arthropoda	Trilobitomorpha		Various skeletal remains and especially whole skeletons of young individuals
			Chelicerata	Merostomata including Gigantostraca = Eurypterid	Chitinous teguments
			Crustacea	Branchiopoda Ostracoda Copepoda Cirripeda Malacostraca	**Calcareous carapaces** and coprolites
			Echinodermata	Crinoidea Asteroidea Ophiuroidea Echinoidea Holothuroidea	Plates and *Lombardia* Radioles and pedicelles Calcareous sclerites
			Hemichordata	including Pterobranchia Graptolithina (= Graptolites)	Carbonised chitinous remains
			Urochordata	Ascidiacea	Calcareous sclerites
	Chordata	Vertebrata	Agnatha	Cephalaspidomorphi including Cyclostomata	Conodonts
			Gnathostomata	'Fish' • Placodermi • Chondrichthyes • Acanthodii • Osteichthyes Amphibia Reptilia Aves Mammalia	Teeth, scales and otoliths Teeth (Rodents)	Bone fragments

BIBLIOGRAPHY

The classification scheme shown here may be supplemented by reference to the following works: D.M. Raup and S.M. Stanley, *Principles of Paleontology* (W.H. Freeman, New York, 1971); R.C. Moore, *Treatise on Invertebrate Paleontology* (University of Kansas Press/Geological Society of America). For detailed study of the relative place of Dinoflagellates, Chrysomonads, Foraminifera, Radiolaria and Heliozoa see N.D. Levine *et al.*, "A Newly revised classification of the Protozoa" (*J. Protozool.*, 27 (1) 37–58).

Part 2
Geological and Palaeobiological Applications of Micropalaeontology

Chapter 12
Microfossils in the Environment of Preservation

Although the fossilization of microorganisms is not fundamentally different from that of larger organisms, it does have several distinctive features.

1. FROM LIVING ORGANISM TO MICROFOSSIL

Protoplasm

When an organism dies, the non-mineralized constituents of the protoplasm are totally destroyed – if oxygen is present – by the action of aerobic bacteria. In this process of putrefaction, the end-products are returned to their place in the mineral world. If, however, the environment is anoxic (i.e. deficient in oxygen), bacterial decomposition remains incomplete. A process of fermentation then leads to the formation of carbonaceous substances and these evolve into kerogene. In this case, the organic matter is partially preserved but, apart from the sporopollenins or 'chitins' in palynomorphs, it is not fossilized in its original form.

Mineralized Tissues

Mineralized tissues may well undergo fossilization unless destroyed or rendered fragile during the lifespan of the organism by the phenomena of autolysis (as is the case with the microspheric forms of foraminifera).

When an organism dies, the first effect of the destruction of the soft parts is to bring the mineralized tissues into contact with the surrounding environment. If the skeletal elements are isolated or are not held together well, they become dispersed. Many macrobiotic elements are thus known only in the form of microfossils (e.g. spicules of sponges, sclerites of ascidians and holothuroids, and conodonts).

Because of their minute size, the skeletons of microorganisms and isolated organs are very little affected by the movements and friction arising from disturbances in the environment. Some sandy sediments contain well-preserved foraminifera and coccoliths alongside the broken shells of molluscs. Microfossils are more often corroded than worn or broken. Corrosion takes two forms. It may have a biological origin, resulting, for example, from ingestion of the microorganism by a predator or attack by a boring alga or fungus. Alternatively, it may be physicochemical, in which case it varies according to the mineralogy (Figs 12.1 and 2) and structural peculiarities of the mineralized tissue.

Aragonite	Skeletons of dasycladalean and udotacean chlorophytes Madreporanan and Hydrocorallian Numerous shells of molluscs (including pteropods) Ascidian sclerites Occasional small foraminifera Fish otoliths
Strongly magnesian calcites $> 7\%$ mol $MgCO_3$ (up to 36%)	Alcyonarian sclerites Calcisponge spicules Cyanophyte sheaths, cell wall of corallinacean rhodophytes All foraminifera with porcelaneous tests Some small and most large foraminifera (e.g nummulitids) with hyaline tests Echinoderm skeletons, serpulid polychaete tubes
Slightly magnesian calcites $< 7\%$ mol $MgCO_3$	Most small foraminifera with hyaline tests including planktonic types Ostracod carapaces Most bryozoans, brachiopods and cirripeds Cell walls and gyrogonites of charophytes Coccoliths

Fig. 12.1 Mineralogical composition and stability during fossilization of some calcareous skeletons. Aragonitic skeletons (or hard parts) are the most fragile: they dissolve or are rapidly transformed into calcite. Calcitic skeletons that are strongly magnesian lose their magnesium and are transformed into calcite without visible modification. Calcitic skeletons with a low magnesium content are stable. In each group, the skeletons are arranged in order of increasing fragility from the bottom to the top

Tests of phaeodarian radiolarians (unknown in the fossil state) Tests of silicoflagellates Frustules of diatoms Tests of polycystine radiolarians Spicules of siliceous sponges

Fig. 12.2 Stability of some siliceous skeletons on fossilization: they are arranged in order of increasing fragility from the bottom to the top

The corrosion of biogenic carbonates, which may lead to total dissolution, is significant in certain continental sedimentary environments (peats) and in shallow marine environments (mangroves, estuarine and littoral muds). Here, the pH may fall to 5 or 4 a few centimetres below the interface between water and sediment. In such environments, small calcareous tests with a high surface/mass ratio are dissolved before larger shells. At pH 5, they disappear within a day.

In oceanic environments, the dissolution of biogenic carbonates is of special importance. Planktonic microorganisms, which live for the most part in surface waters (Fig. 12.3), sink to the bottom after their death. The tests of foraminifera and radiolarians settle after a few days but coccoliths and frustules of small diatoms take several years. On the surface of the sediments, the tests of foraminifera undergo corrosion, which varies according to the

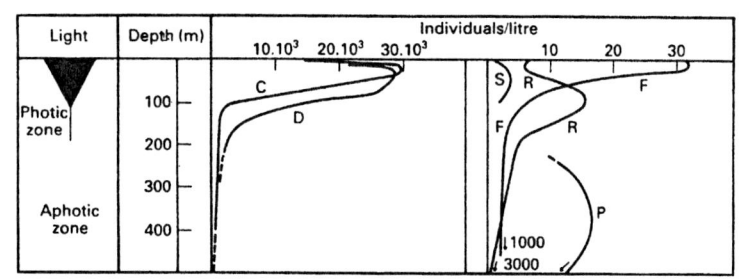

Fig. 12.3 Average quantitative distribution of individuals per litre of water for some Recent planktonic microorganisms as a function of depth and light: C = coccolithophores; D = diatoms; F = foraminifera; P = pteropods; R = radiolarians; silicoflagellates. The abundance of producer phytoplankton is in contrast to the small number of consumer zooplankton

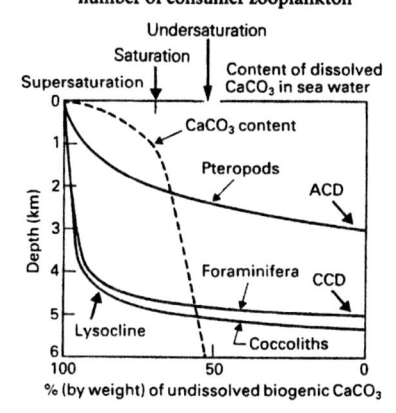

Fig. 12.4 Average quantitative distribution of individuals per litre of water for some Recent planktonic microorganisms as a function of depth and light: C = coccolithophores; D = diatoms; F = foraminifera; P = pteropods; R = radiolarians; s = silicoflagellates. The abundance of producer phytoplankton is in contrast to the small number of consumer zooplankton

species and to depth (Fig. 12.4). The effect is not significant until a depth of around 4000 m is reached. Below this level, which is called the **lysocline**, the rate of dissolution increases sharply, together with a growing undersaturation of the water in carbonates. Tests are not preserved beyond the **calcite compensation depth** (CCD), which is situated between 4000 and 5000 m.

Coccoliths continue to slightly greater depths. They are more resistant because they are frequently enclosed in organic matter: copepods, which may consume up to 1200 coccolithophores a day, eject coccoliths almost intact in some 30 to 40 faecal pellets every day. The aragonite of pteropod shells is much more fragile, disappearing before a depth of 3000 m. Lysoclines and the depths for the disappearances of each of the groups of microorganisms vary

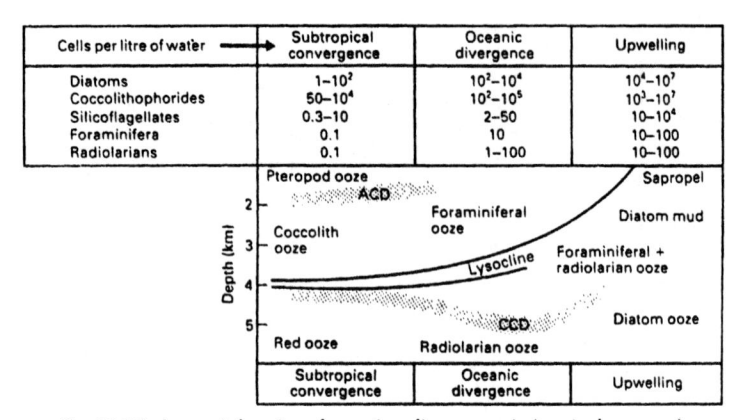

Cells per litre of water	Subtropical convergence	Oceanic divergence	Upwelling
Diatoms	$1-10^2$	10^2-10^4	10^4-10^7
Coccolithophorides	$50-10^4$	10^2-10^5	10^3-10^7
Silicoflagellates	$0.3-10$	$2-50$	$10-10^4$
Foraminifera	0.1	10	$10-100$
Radiolarians	0.1	$1-100$	$10-100$

Fig. 12.5 Bathymetric layering of oceanic sediments, variations in the aragonite compensation depth (ACD) and the calcite compensation depth (CCD), and variations in the lysocline, in the Pacific Ocean as a function of planktonic productivity. After Berger (1976, fig. 29.2)

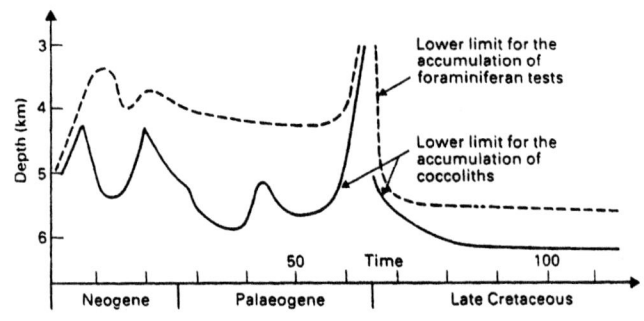

Fig. 12.6 Fluctuations in the depth at which tests of planktonic foraminifera and coccoliths disappeared during the Cenozoic and the late Cretaceous. After Hay (1970, fig. 4)

according to planktonic productivity: the higher the supply of biogenic carbonates, the deeper the level (Fig. 12.5). The limits have evolved considerably throughout geological time (Fig. 12.6).

Siliceous tests similarly undergo a considerable degree of corrosion (Fig. 12.7) in surface waters that are undersaturated in silica. Beyond a depth of 1000 m, they are subject only to moderate and progressive attack.

In sediments, microfossils, like larger remains, do not always have a chemical composition identical to that of the original mineralized tissue. The transformation usually takes place early in sediments subject to diagenic (or diagenetic) physicochemical conditions that are similar to those of the living organism. Transformation may take place without a change in chemical

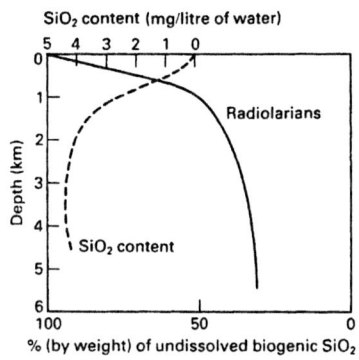

Fig. 12.7 Rate of dissolution of siliceous skeletons of radiolarians in the tropical Pacific Ocean as a function of depth and the SiO_2 content of the water.

composition, as is the case when aragonite is turned into calcite, or opal into chalcedony. It may also be accompanied by a loss of ions, particularly Mg^{2+}, with the transformation of magnesian calcites into pure calcite. The organic materials included in the mineralized tissues are modified but, being protected, do not totally disappear. Often the chemical composition differs and the fossil may then be of calcium phosphate, chlorite, goethite, pyrite, glauconite or carbonaceous materials, etc. The transformation takes place according to two distinct mechanisms – either by epigeny (the substitution of the biogenic material ion by ion) or by the filling of the imprint left by the total dissolution of the organism within the consolidated sediments. In the former case, the initial form and microstructure remain although frequently altered; in the latter, they disappear.

Balance

During fossilization, potential microfossils are not individually favoured in comparison to larger species. Their large initial numbers, however, are a favourable factor. On 10 cm^2 of neritic sea-bed there is an average of 50 to 200 benthic foraminifera. A litre of surface water from the ocean will contain a few pteropods and planktonic foraminifera, a dozen silicoflagellates and radiolarians, and thousands of diatoms, peridinian dinoflagellates and coccolithophores. Although these values may be multiplied a thousandfold at a particular place and time, they nevertheless represent permanent densities. Individuals reproduce, die and are replaced after a lapse of time that is variable but, like their lifespan, short. Thus, the 10 cm^2 of sea-bed that shelters 100 foraminifera at any one time, will produce an annual total of 400 to 600 individuals.

What can be said of spores and pollens that are emitted in unimaginable quantities? A single pine tree produces more than 12 000 million grains of pollen per year ('sulphur rains'). The pine may be a particularly prolific case but, although angiosperms do not produce as much pollen they are not far behind.

Micropalichnology

Palichnology is the study of the traces of biological activity left by organisms. When these are microscopic, they fall within the domain of micropalaeontology. Apart from stromatolites, which will be discussed in Chapter 14, the following examples can be cited:

- Coprolites (faecal pellets) produced by various benthic organisms such as polychaete annelids, gastropods and crustaceans. They are often calcified or phosphatized on diagenesis. Some, like *Favreina* (Fig. 12.8), known from the Triassic onwards, are occasionally abundant in limestones (pelmicrites) of lagoonal origin.
- Fine, branching canaliculi extending through calcareous substrates, tests or bioclasts. They are due to the action of thallophytes: bacteria, algae and/or fungi (Fig. 12.9).
- Conical perforations (20 to 40 μm), which are probably the work of predators (e.g. gastropods and nematodes), passing through the walls of foraminiferal tests (Fig. 12.9).

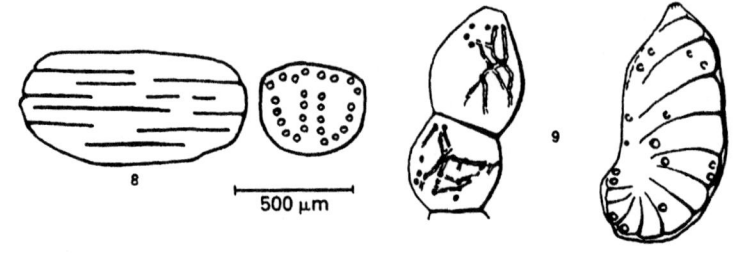

Fig. 12.8 Coprolites (*Favreina*) traversed by fine canaliculi: axial section (left); transverse section (right) (×40). After Brönnimann (1976, figs 2 and 4)

Fig. 12.9 Perforations in the tests of fossil foraminifera, probably due to thallophytes (left) or to gastropods (right) (×40). After Pozaryska (1957, figs 5 and 19)

2. DEPOSITION

Microfossils are not distributed haphazardly in sediments but are often concentrated in deposits. There are, moreover, qualitative and quantitative differences between the initial living association (**biocoenose**) and the assemblage of microfossils (**taphocoenose**) that results. The modifications are due to a number of phenomena.

Selective dissolution

Dissolution in water and on the ocean bed varies according to the types of organism and to the mineralogical nature of the skeletons. This results in a bathymetric stratification of oceanic muds (Figs 12.5, 14.1 and 14.2).

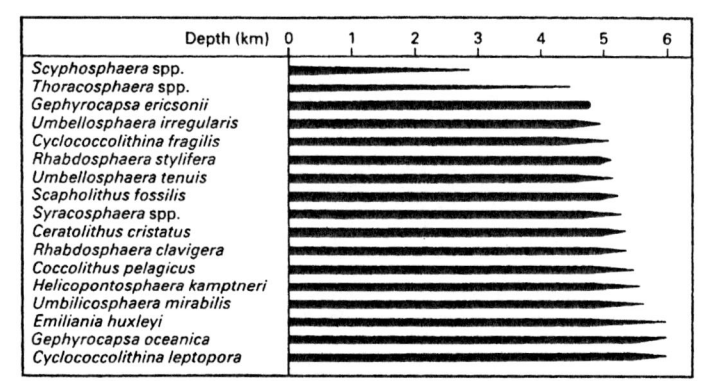

Fig. 12.10 Depth distribution of some species of coccoliths in Recent marine sediments. Mainly after Roth and Berger (1975, text fig. 10)

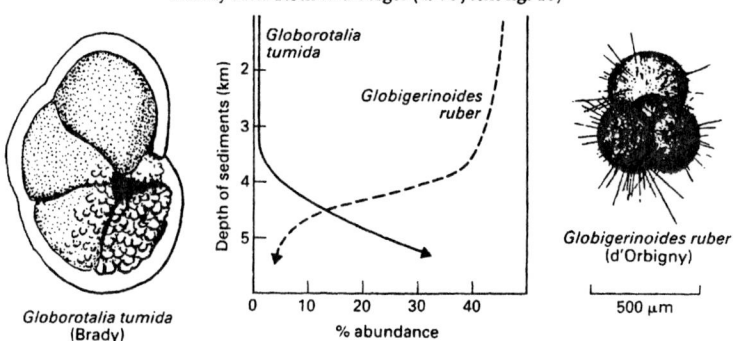

Fig. 12.11 Changes in abundance of two species of planktonic foraminifera in Recent oceanic sediments as a function of depth. The two species are seen at the same magnification (×30). *G. tumida*, has a large robust test with a thick carina; *G. ruber* has a small thin test covered with fine spines. After Heezen & Ruddiman (1967, fig. 8)

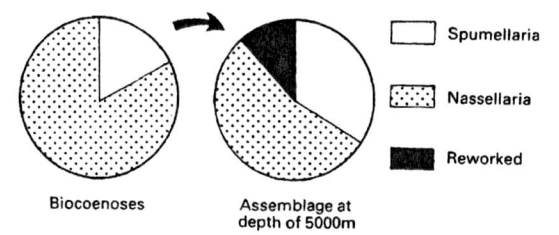

Fig. 12.12 Comparison between an association of living planktonic radiolarians and an assemblage of corroded and reworked radiolarian tests from Recent sediments in the equatorial Pacific Ocean. After Petrushevskaya (1976, fig. 21.6)

140 PART 2: GEOLOGICAL AND PALAEOBIOLOGICAL APPLICATIONS

Similarly, dissolution is selective at the level of species within each group: those with delicate tests disappear before those that are more robust, the latter becoming increasingly abundant with depth (Figs 12.10 and 12.11). The assemblage of dead individuals (**thanatocoenoses**) has a numerical and specific composition, which is different from that of the biocoenoses from which it arose (Figs. 12.10–12).

Modifications such as these are also known in continental and in neritic marine environments. The depend, however, on factors other than depth. In the supratidal domain, for example, where pH decreases during the night, the calcareous tests of benthic foraminifera dissolve whereas those of chitinoid cement agglutinates remain unaffected. Pollens also display a variable resistance. As a rule, those of Coniferae, Caryophyllaceae and Compositae survive whereas those of the Gramineae are rapidly destroyed.

Natural Displacement and Bioturbation

Although the skeletons of planktonic microorganisms concentrated in sediments appear allochthonous, graded bedding, sorting and preferential orientations distinguishable in the microfacies, all indicate that the most microfossils, whatever their mode of life, were moved before burial. Autochthony is impossible to prove except in rare cases where the fossilized skeleton of a fixed organism has remained attached to a support that has not moved. Given the variability in amount of postmortem displacement, micropalaeontological deposits are, for the most part, allochthonous. Confirmation of this has come from palynologists who have, on many occasions, underlined the role of wind in the dispersion of pollens (Fig. 12.13) just as sedimentologists have described the way that currents transport shells (Fig. 12.14). The smaller the shell, the more easily it is moved. Thus, thanatocoenoses tend to be uniform and to contain only the most resistant remains. Under the action of currents, a shallow marine area, a delta or estuary produces thanatocoenoses that are quite different from biocoenoses. In lakes and in deeper sea waters, the thanatocoenoses are less distorted. Finally, turbidity currents carry neritic sediments into the oceanic domain. This explains the alternation of nummuli-

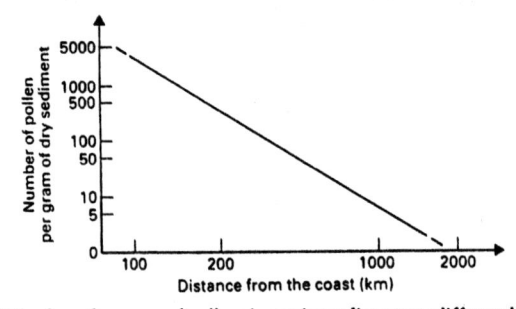

Fig. 12.13 Number of spores and pollens in marine sediments at different localities in the Atlantic, Indian and Pacific oceans, plotted against distance from the nearest coast. After Groot & Groot (1971, fig. 37.2)

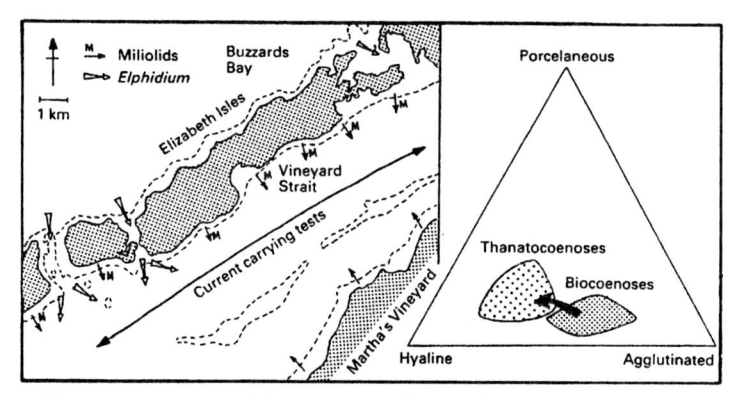

Fig. 12.14 Biocoenose and thanatocoenose of benthic foraminifera in the Vineyard Strait (east coast of the USA between Boston and New York). In the axis of the strait, the biocoenose have 20 to 50% of fixed forms (e.g. *Eggerella* and *Trochammina*) and some *Rosalina*. At the same points, the thanatocoenoses show a lower percentage (4 to 30%) of fixed forms mixed with miliolids and *Elphidium*. The latter live in littoral areas and are carried by tidal currents. After Murray (1976, figs 13 and 14)

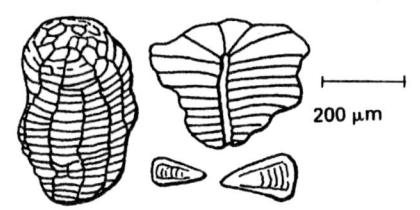

Fig. 12.15 *Microcodium:* left, complete 'thallus'; right, axial section (above) and isolated prisms (below) (×34). After J. Sigal (1957, figs 1, 3 and 5)

tic limestones with planktonic microfossil marls in Palaeogene flysches.

After deposition, the skeletons in unconsolidated sediments continue to be turned over and homogenized by the action of burrowing animals; these penetrate up to varying depths (40 cm in neritic marine environments and 10 cm on the ocean bed).

With the consolidation of the sediments, the activity of the burrowing organisms is interrupted. The hardened rock may still be penetrated, however, by boring organisms such as *Microcodium*, which is formed from small prismatic elements arranged like the grains on a corn-cob (Fig. 12.15). These microfossils may result from the calcification, during continental conditions, of mycorrhiza inhabiting soils over limestone substrates. *Microcodium* is not known before the Palaeocene. Similar forms have been found in Recent soils.

Reworking

Just as microfossils may be moved in space before burial, so they can be displaced in the course of time. In this case, the transfer is described as reworking. This can vary in amount, but it goes further than the action of burrowing organisms, which can also mix thanatocoenoses of different ages.

Reworked organisms that have been repositioned underwent an initial sedimentation and fossilization. These processes were followed, after a certain period of time, by the removal of the original deposit from continental or submarine outcrops, transport, and then, finally, re-sedimentation. Because of their minute size and great abundance, small fossils are more subject to reworking than large ones. Foraminifera, coccoliths and other microfossils are common in the sediments of both recent (Fig. 12.12) and ancient times. In the Paris Basin, the Eocene levels lie directly over an eroded chalk surface and contain microfossils, foraminifera and coccoliths that have been removed from the chalk. These are often better preserved and more numerous than the Eocene microfossils themselves. The greater the difference in age between reworked elements and those contemporaneous with the sedimentation, the easier it is to recognize the reworking. Sometimes, however, it can only be suspected.

The assemblages of microfossils in deposits result, therefore, from a series of selective mechanisms and from remains that may or may not be contemporaneous with the sedimentation.

3. MICROFOSSILS IN DEPOSITS

When a deposit becomes exposed, it is destroyed or dispersed by the agents of erosion. Human activity is also a factor in this destruction, which is far from negligible. However, although the reserves of large fossils have been brought almost to exhaustion by the activities of collectors and commercial dealers, deposits of microfossils remain virtually untouched. They are indeed so numerous and so rich that they can be considered inexhaustible.

After sedimentation and diagenesis, fossils of every size are subject to deformation and crushing as they are compacted. Large foraminifera are sometimes pierced by quartzes or truncated by stylolitic surfaces. Such damage is relatively small compared with the action of metamorphism on buried microfossils. For a long time, it was thought that identifiable fossils could not be found in sediments that had undergone major metamorphism. This is generally true for large fossils. In the case of calcareous and siliceous microfossils, thin sections show their outline to be blurred, blending into a cement that has largely recrystallized. However, microfossils have been found in a limestone block encased in basalt, radiolarians have been seen included in albite crystals and acritarchs are known from authigenic quartzes. These discoveries have prompted new research, which has proved fruitful. Many metamorphic rocks – some of the greenschist facies and fewer of the blue schist facies – have, after suitable preparation, yielded determinable radiolarians, palynomorphs and conodonts.

These organic remains, which persist when all others have disappeared, are nevertheless altered. The profound chemical and mineralogical modifications undergone by sporopollenin and phosphatic microfossils can be seen in changes of colour that are progressive, cumulative and irreversible. They depend on the depth of burial (i.e. the temperatures reached at these depths) and on the period of time that they are subject to these conditions. Palynomorphs pass through several colorimetric stages (five or ten according to specialists) at varying rates: spores and pollens first, followed by dinocysts and finally chitinozoans. Originally pale yellow, they become orange to red, brown and eventually black. The growing opacity of the walls results from a loss of hydrogen and oxygen and from enrichment in carbon (carbonization). The process is accompanied by a gradual obliteration of the morphological characteristics. In the same way, conodonts pass through colorimetric stages categorized in the Conodont Alteration Index (CAI). There are eight of these: CAI 1, pale yellow; CAI 2–4, deepening shades of brown; CAI 5, black; CAI 6, brown; CAI 7, opaque white; CAI 8, transparent white. The opacification of the first stages is the consequence of carbonization of the organic matter dispersed between the crystals of apatite. The lightening that takes place in the latter stages is explained by the progressive disappearance of carbon, which eventually becomes complete.

These colorimetric stages have been standardized in the laboratory so that the colour changes of palynomorphs (Fig. 12.16) and conodonts (Fig. 12.17) can now be used as palaeothermometers. Provided that there are no tectonic complications, they can also be used to assess the thickness of the overlying sediments.

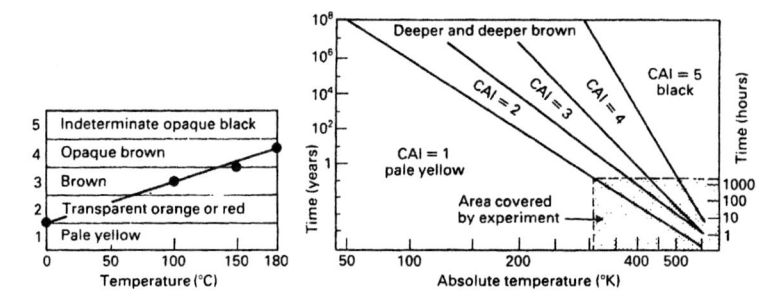

Fig. 12.16 Colorimetric evolution of palynomorphs (spores and dinocysts) exposed to temperatures of 100 to 180°C for 300 and 1500 hours. After Correia (1967, fig. 10)

Fig. 12.17 Colorimetric evolution of conodonts as a function of temperature and duration of heating. Curves established experimentally and extrapolated for geological time. CAI 8 is obtained by heating for 4 hours at 950°C. After Epstein, Epstein & Harris (1977, figs 3 and 9)

CONCLUSION

By virtue of their initial abundance and their better distribution in the various environments of fossilization, microorganisms are more likely to leave traces in sediments than larger organisms. Moreover, certain groups of microfossils are resistant to the effects of metamorphism to some degree. A rock that to the naked eye appears devoid of organic remains may well contain a multitude of microfossils. Their juxtaposition or taphocoenosis, however, differs from the initial biocoenosis in content and completeness. These highly significant characteristics must be taken into account when microfossils are used for purposes of palaeobiology, stratigraphy or palaeogeography.

BIBLIOGRAPHY

Fossilization and the emplacement of deposits are almost invariably treated with macrofossils as the point of departure. On this subject see C. Babin, *Eléments de Paléontologie*, pp. 14–29 and 96–136, (Colin, Paris, 1971) and J. Roger, *Paléontologie générale*, pp. 163–201 (Masson, Paris, 1974).

There is an abundant literature covering selective dissolution and the sedimentation of microplankton skeletons in an oceanic environment: B.M. Funnel & W.R. Riedel (eds), *The Micropaleontology of Oceans* (Cambridge University Press, 1971); W.W. Sliter, A.W.H. Bé & W.H. Berger, *Dissolution of Deep-sea Carbonates (Cushman Foundation for Foraminiferal Research*, Special Publication 13, 1975); W.H. Berger, 'Biogenous deep-sea sediments, in J.P. Riley & R. Chester (eds), *Chemical Oceanography*, vol.5, pp. 265–388 (Academic Press, London, 1976); and A.T.S. Ramsay, *Oceanic Micropalaeontology*, vol. 2, chap. 12 (Academic Press, New York, 1977).

The transition from biocoenoses to taphocoenoses is described and illustrated with examples in J.W. Murray in *Foraminifera*, R.M. Hedley & C.G. Adams (eds), vol. 2, pp. 45–110 (Academic Press, London, 1976).

Little has been said about the action of metamorphism on microfossils except in A.G. Epstein, J.B. Epstein & L.D. Harris, *Professional Paper of the US Geological Survey* 995 (1977; fine colour photographs), and by palynologists and the petroleum geologists such as B.S. Cooper in G.D. Hobson (ed.), *Nouveaux Aspects de la Géologie du Pétrole*, pp. 51–59 (Editions SCM, Paris, 1980).

Chapter 13

Microfossils – The Key to Biological Problems

Although it is a truism to recall that microfossils are the remains of living organisms, it is necessary, paradoxically, to emphasize that their study as such has been, and still is, frequently neglected. One reason for this neglect lies in the relative scarcity of biological studies on living representatives of the groups from which microfossils derive. Coccolithophores and silicoflagellates were discovered in sediments before living examples had been found. The dimorphism of foraminifera was recognized by micropalaeontologists 40 years before it was explained by biologists. This situation goes back to the last century and it has not improved since. The significance of dinocysts and of numerous hystrichospheres was demonstrated in 1961 through the examination of fossils, before being established in living organisms.

Some microfossils are the only traces remaining of life-forms that have no equivalent in nature today. Others, which are related to groups that exist at present, have characteristics that are no longer possessed by their modern descendants. As their morphology is known, it is now necessary to reconstruct their mode of existence.

Microfossils also illuminate many biological problems, in particular the notion of species and speciation, evolutionary trends, and the stages of biogenesis.

1. FROM ECOLOGY TO PALAEOECOLOGY

Basic Concepts

In order to grow and reproduce, an organism requires particular conditions that are found in limited areas called **biotopes**. Ecology involves the study of such conditions. Some of these conditions, being physicochemical, are abiotic. For example, aquatic organisms are affected by the following conditions:

- The temperature – its average value (−2 to + 27°C for the oceans, and up to 35°C in some enclosed seas) and its extremes.
- The ionic content, the sum of which constitutes the salinity (S) expressed in parts per mille (33 to 39‰ for open seas), together with the content of dissolved gases and trace elements.
- The turbulence, due to currents and waves, which separates calm areas of low energy from rough areas of high energy.
- The turbidity or transparency, which, as a function of the intensity of sunlight, determines the depth of the photic layer.
- The depth.
- The nature of the substrate (indurated or loose), its grain-size distribution and stability, and the rate of sedimentation.
- Certain geological phenomena such as volcanism, which is responsible for the high development of microorganisms with siliceous skeletons (relationship between basalts or pyroclastic tuffs and diatomites).

To these conditions must be added a number of biotic factors such as:

- The supply of nutrients, which governs the primary production of phytoplankton and thus the secondary production of predators.
- Competition between species and the ratio of producers to predators.

Habitats are numerous and modes of existence equally so. **Ethology**, which studies them, divides the populations of aquatic environments into two main groups: **pelagic** and **benthic** (Fig. 13.1).

Species, Ecomorphs and Ecological Parameters

Optimal and lethal values of ecological factors for different species have been determined by laboratory experiment. It must be remembered, nevertheless, that:

- It is impossible, at present, to culture large numbers of microorganisms, particularly pelagic ones.
- The behaviour of individuals raised *in vitro* is sometimes different from that of individuals in nature; this is explained in part by complex interference phenomena, or by compensation between factors.

The ecology of species is better established by direct examination of populations in natural environments. The distribution of organisms is a function of ecological parameters (Figs 13.2 and 3). Two types can be distinguished: stenobiota, which are sensitive to even minute variations in the environment, and eurybiota, which can tolerate considerable variations.

Organisms with identical modes of life are often somewhat similar, for example, planktonic foraminifera and certain ostracods. However, although identical physicochemical factors result in morphological convergence, ecological differences work in the opposite direction to modify the characteristics of representatives of the same species or genus. These variants are said to be **ecomorphs** or **ecotypes**. Thus, for example, the euryhaline ostracod, *Cyprideis torosa* (Jones), is represented in marine and hyposaline ($S \geqslant 5‰$) environments by individuals with a smooth carapace. In lacustrine environments, however, or in others of very low salinity, the populations have a tuberculate

MICROFOSSILS – THE KEY TO BIOLOGICAL PROBLEMS

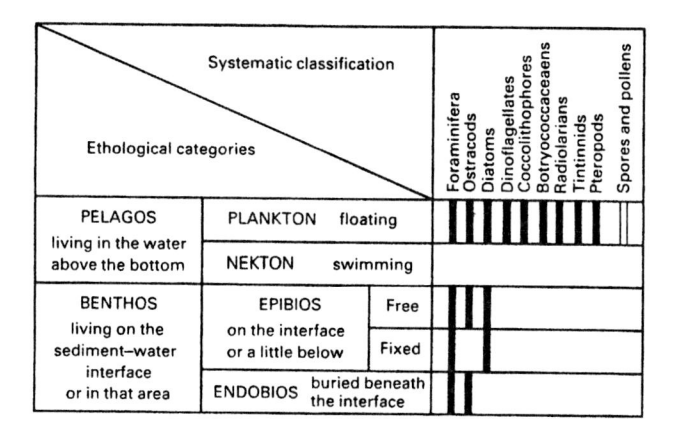

Ethological categories	Systematic classification		Foraminifera	Ostracods	Diatoms	Dinoflagellates	Coccolithophores	Botryococcaceaens	Radiolarians	Tintinnids	Pteropods	Spores and pollens
PELAGOS living in the water above the bottom	PLANKTON floating											
	NEKTON swimming											
BENTHOS living on the sediment–water interface or in that area	EPIBIOS on the interface or a little below	Free										
		Fixed										
	ENDOBIOS buried beneath the interface											

Fig. 13.1 Ethology of some groups of living microorganisms with fossil counterparts. Spores and pollens are not planktonic microbiota but because of their mode of dispersion they can be treated in the same way

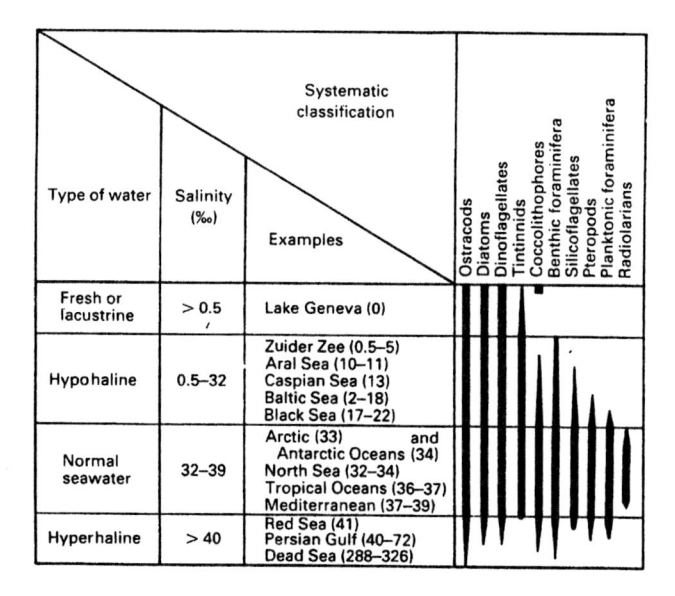

Type of water	Salinity (‰)	Examples	Ostracods	Diatoms	Dinoflagellates	Tintinnids	Coccolithophores	Benthic foraminifera	Silicoflagellates	Pteropods	Planktonic foraminifera	Radiolarians
Fresh or lacustrine	> 0.5	Lake Geneva (0)										
Hypohaline	0.5–32	Zuider Zee (0.5–5) Aral Sea (10–11) Caspian Sea (13) Baltic Sea (2–18) Black Sea (17–22)										
Normal seawater	32–39	Arctic (33) and Antarctic Oceans (34) North Sea (32–34) Tropical Oceans (36–37) Mediterranean (37–39)										
Hyperhaline	> 40	Red Sea (41) Persian Gulf (40–72) Dead Sea (288–326)										

Fig. 13.2 Simplified classification of aquatic environments and salinity ranges for some groups of living microorganisms.

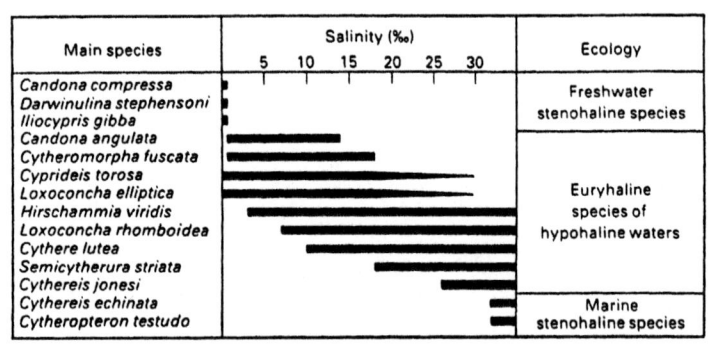

Main species	Salinity (‰) 5 10 15 20 25 30	Ecology
Candona compressa		Freshwater stenohaline species
Darwinulina stephensoni		
Iliocypris gibba		
Candona angulata	▬▬▬▬	Euryhaline species of hypohaline waters
Cytheromorpha fuscata	▬▬▬▬▬	
Cyprideis torosa	▬▬▬▬▬▬▬	
Loxoconcha elliptica	▬▬▬▬▬▬▬	
Hirschammia viridis	▬▬▬▬▬▬	
Loxoconcha rhomboidea	▬▬▬▬▬▬	
Cythere lutea	▬▬▬▬▬	
Semicytherura striata	▬▬▬▬	
Cythereis jonesi	▬▬▬	
Cythereis echinata	▬	Marine stenohaline species
Cytheropteron testudo	▬	

Fig. 13.3 Salinity ranges of some living ostracod species on the coast of the North Sea and Zuiderzee. After Wagner (1957, figs 2 and 3)

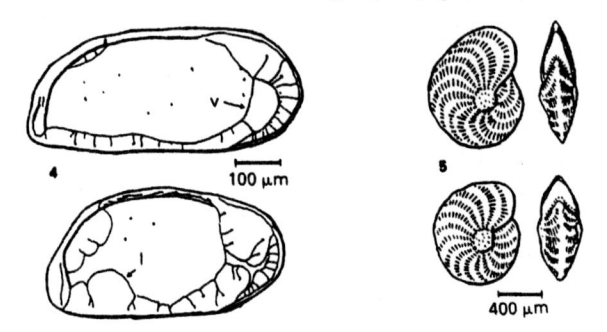

Fig. 13.4 Internal morphology of the valves of individuals of the same ostracod genus (*Krithe*) as a function of the dissolved oxygen content of water. Above, enclosed waters (≤4 ml O_2/ litre of water); → anterior vestibule (v) is wide and internal ventral lamella is small. Below, well-oxygenated waters (≥ 6.5 ml O_2/ litre of water); → vestibule is small and lamella (l) highly developed (×75). After Peypouquet (1975, pl. 1, fig. 2 and pl. 2, fig. 8)

Fig. 13.5 Ecomorphs of the same species of foraminifera: *Elphidium crispum* (Linne) as a function of food supply. Above, abundant supply; below, reduced. In the latter case, the test is more robust (×18). After Myers (1943, fig. 4A, B)

carapace. Further modifications are linked with the content of dissolved oxygen in the environment (Fig. 13.4), with the concentration of phosphorus and silicon, and so on. Foraminifera provide good examples of ecomorphs linked with food supply (Fig. 13.5) or temperature (Fig. 13.6).

The chemical composition of the test depends, in part, on the temperature. In calcitic foraminifera, for example, the higher the temperature, the higher the magnesium content. Moreover – as will be seen later – the isotope ratios of oxygen, calcite and biogenic silica vary with the temperature of the environment.

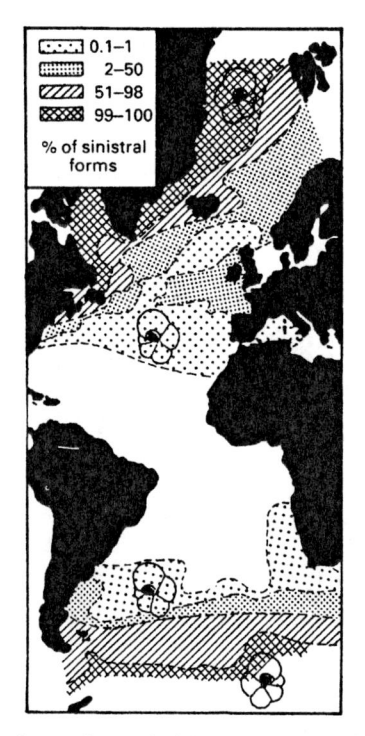

Fig. 13.6 Percentage of sinistral tests of *Globoquadrina pachyderma* (Ehr.) in surface waters (< 10 m) of the Atlantic Ocean. Sinistral forms predominate in cold waters and dextral ones in warmer waters. The dominance of dextral populations along the coasts of northern Europe indicates the presence of the Gulf Stream coming from the south-west. After Bé & Tolderlund (1971, fig. 6.6)

In nature, every ecological factor has a role. Some are more significant than others but the problem of interference makes it difficult to assess them individually. Depth, for example, is in itself probably less a determinant factor than the thermal, dynamic, chemical and photic modifications that depend on it.

Biocoenoses and Biotopes

Although there are interferences between abiotic factors, there are also biotic interactions. Each species has an effect on the others and forms an integral part of a community or biocoenose. This is linked to a particular biotope in such a way that the overall ecosystem or environment forms a whole from which the parts cannot be dissociated.

Marine planktonic biocoenoses cover vast areas with biotopes consisting of

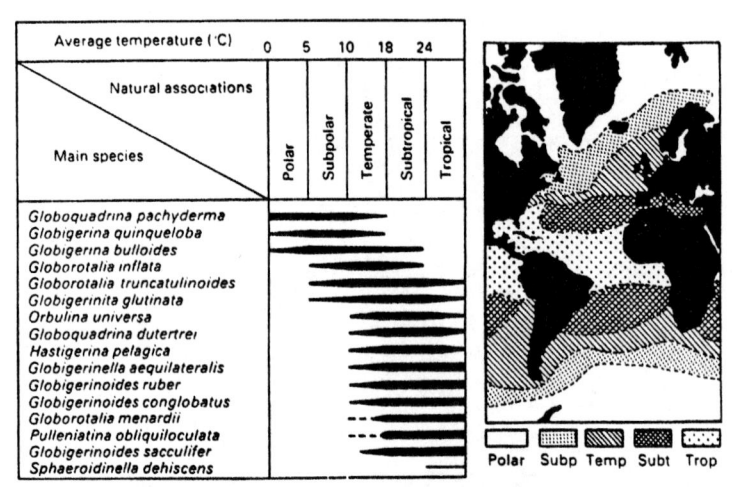

Fig. 13.7 Natural associations and bioprovinces of planktonic foraminifera in the Atlantic Ocean. The arrangement does not follow latitudes exactly because the Gulf Stream brings temperate waters towards the north along the coasts of Europe. After Bé & Tolderlund (1971, fig. 6.2)

homogeneous ocean masses. The surface waters, which are richest in plankton, are divided into **provinces** according to temperature in a bipolar and roughly latitudinal arrangement (Fig. 13.7). In addition, for organisms that are not bound to the photic layer, there is a tiered arrangement according to depth with tropical submergence (Fig. 13.8). The circulation of water masses produces corresponding biocoenoses. The surface currents disturb the latitudinal arrangement, while upwellings bring deep associations to the surface.

Benthic biotopes include both the substrate and a certain thickness of the overlying water. In the neritic marine domain these biotopes are varied and localized. This diversity is not found in the ocean depths. At depths below 500 to 1000 m, the water is dark with low ($\leq$ 8 to 10 °C) but stable temperature, low salinity ($\leq 35‰$) and high pressure. As a result, the cold water circulation supports a special, cosmopolitan fauna. The **psychrosphere** (from the Greek *psychro* = cold) results from the spreading out at depth of cold polar waters of low salinity. It stands in contrast to the surface waters of the **thermosphere** where the temperature is higher but unstable.

One of the most influential factors in aquatic environments is **salinity** (Fig. 13.9). The greatest diversity in species and abundance in numbers of individuals is found where the salinity is zero (lacustrine environments) or normal and stable (marine environments at $32‰ \leq S \leq 39‰$). By contrast, environments where the salinity is low (**hypohaline**) or high (**hyperhaline**) contain dense oligo- or monospecific populations in which each group is represented by a few species that are able to multiply greatly because of the absence of competition between species that is characteristic of more favourable environments.

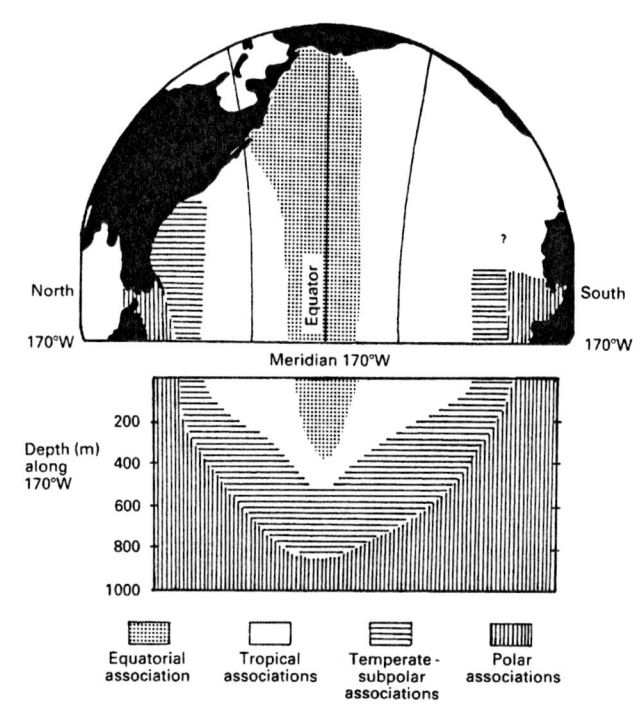

Fig. 13.8 Latitudinal distributions of associations of Recent polycystine radiolarians, in the surface waters of the Pacific Ocean (above) and tiered according to depth along the meridian 170°W (below). After Casey (1971, fig. 7.1)

The interaction of ecological factors conditions the distribution of organisms (Fig. 13.10). In their turn, organisms may modify the environment. For example, the nummulitid foraminifera, by creating barriers parallel to the coast, alter regional topography and the distribution of neighbouring habitats (Fig. 16.1).

Palaeoecological Reconstruction

In nature today, variations in ecological factors entail qualitative and quantitative modifications in biocoenoses. It follows, therefore, that the characteristic features of taphocoenoses should make it possible (to some extent at least) to reconstruct the modes of existence of organisms in the past.

Ecological comparisons are validated by **biosystematic descent**. In this way, it can be considered that the Palaeogene *Nummulites* lived in environments similar to those of Recent *Operculina* (Fig. 3.7). However, although frequently used, the postulate that fossil specimens of a given species, genus or family had

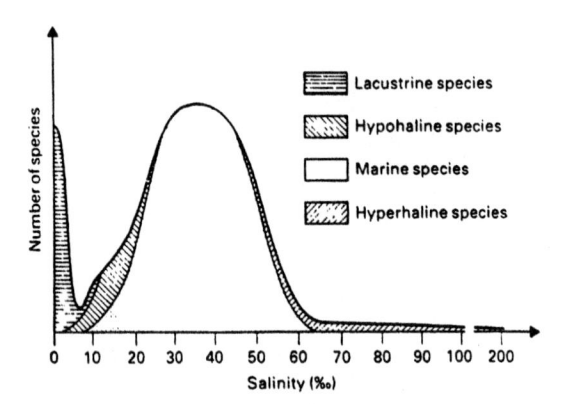

152   PART 2: GEOLOGICAL AND PALAEOBIOLOGICAL APPLICATIONS

Fig. 13.9 Relative abundance of species living in modern habitats of different salinity. The variations indicated by the curve are valid for many groups of aquatic organisms but the absolute number of species varies for each of the groups. After Kinne (1970, fig. 4.73)

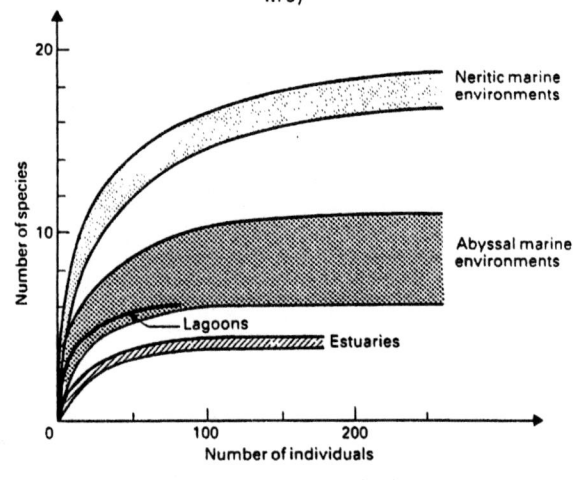

Fig. 13.10 Relationship between the quantity of ostracods and the number of species living in some modern habitats. After Benson (1973, fig. 8)

an ecology comparable to that of their living representatives, cannot always be accepted without question. Species of the same genus, or genera of the same family, may live in very different environments. Moreover, groups, and even genera, may, despite the absence of significant morphological modifications, display an ecological evolution with a slow migration towards the sanctuary of other environments – hypohaline and lacustrine for cyanophytes, psychrospheric for various foraminifera (Nodosariidae) and certain ostracods.

A second method of approach is through **morphofunctional analysis**, based on the probability that organisms of identical form had an identical mode of existence. Although full of promise, this method is not yet widely applied. It has, however, demonstrated that some fossil foraminifera were planktonic, and that others, deformed by fixation, would have required some sort of support, e.g., a hard substrate, another organism preserved or not (e.g. seaweed). Similarly, the method provides justification for the ecological comparison of the alveolinid and fusulinid foraminifera. Such interpretations, however, become more controversial when the fossils in question are very old or the groups have entirely disappeared.

Even in the latter case, palaeoecologists are not entirely without resources. Their method then is to take into account the whole of the taphocoenose. Thus, it can be seen that *Nannoconus* and calpionellids, associated in the same sediments with coccoliths, radiolarians and ammonites, belong, with virtual certainty, to the oceanic plankton. In this particular case, the problematic microfossils are mixed with the representatives of groups, some of which are still extant. In Palaeozoic taphocoenoses, however, the conclusions are less certain as they differ considerably from present-day populations.

Starting out, therefore, from the initial data provided by biologists, palaeontologists proceed through a long and difficult struggle with clues from systematics, morphology and taphonomy until they succeed in reconstructing, step by step, the ecology and ethology. It is, of course, understood that the degree of certainty diminishes rapidly the further back the investigation goes in time.

2. SPECIES AND SPECIATION

Systematics is the study of the classification of living things, the different categories or taxons and their reciprocal relationships. The fundamental taxon is the species and the demarcation of its limits has been, and still remains, a difficult problem for both biologists and palaeontologists.

Typological and Biological Concepts of the Species

More than two centuries ago, the Swedish naturalist, Linnaeus, defined the species as a collection of individuals, identical or almost identical to an arbitrarily chosen reference (or type) individual. This typological conception, born of a view of nature as fixed, led to the creation of more and more species, each one itself being more and more restricted. Any individual that differed however slightly from the type was considered as a new species with the result that there was an enormous multiplication of taxons, particularly for foraminifera. Gradually, the species lost all biological significance and was scarcely distinguishable from the individual.

For biologists, the fundamental criterion for species is now interbreeding, the characteristic that ensures the continuous circulation of the genetic heritage among individuals and that, in consequence, maintains the relative stability of their morphological characteristics. As palaeontologists are unable to make direct use of the interbreeding criterion, they have resorted to its immediate consequence – morphological variability (Fig. 13.11). Given that the observable characteristics were innumerable, the choice for quantitative assessment fell on those that were most easily measurable. In any collection of present-day individuals or fossils, variability is indicated for the various parameters:

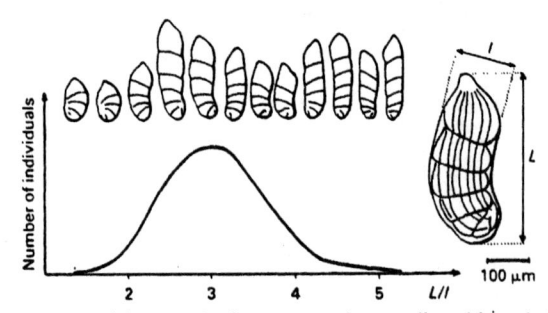

Fig. 13.11 Variability in an isochronous population collected from the lower Cretaceous (Albian) of the Paris Basin. The frequency curve for the ratio L/l is unimodal. All intermediate forms are found between short squat individuals and long thin ones, their outlines being sketched above the curve. The population belongs to a single species of foraminifera: *Marginulina acuticostata* Reuss. After Magniez-Jannin (1975, figs 52 and 53)

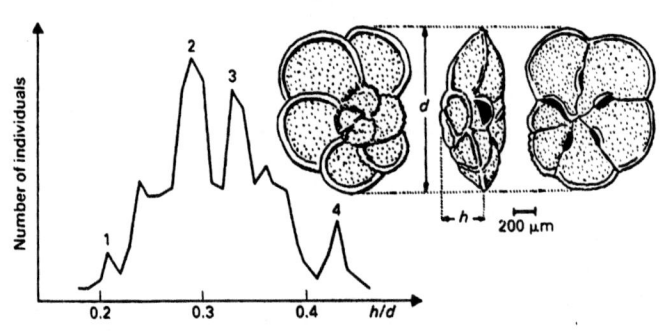

Fig. 13.12 Multimodal variation in the ratio of the height (h) of the spiral curve to the diameter (d) of the test of an isochronous collection of *Rotalipora* collected from the lower Cretaceous (Albian) of the Breggia (Switzerland). The frequency curve contains several peaks indicating the heterogeneity of the population. The determination of the different individuals measured on the basis of characteristics other than the h/d ratio (e.g. number of chambers in last spiral, number of sutural apertures, number of keeled chambers, etc), indicates that there are at least four species of foraminifera, one corresponding to each peak. 1, *R. subticinensis* (Gandolfi); 2, *R.* cf. *appenninica*, an ill-defined group possessing characteristics of the preceding species and, at the same time, features of the species *R. appenninica* Renz which appears later; 3, *R. ticinensis* (Gandolfi); 4, *R. multiloculata* (Morrow). Not all the peaks correspond to the maxima of concentration for species. Some merely indicate interference zones between species. After Caron (1967, fig. 8)

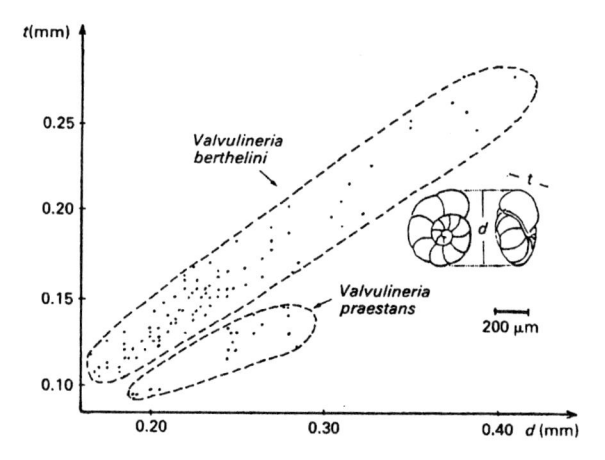

Fig. 13.13 Dispersion diagram for the ratio of thickness (t) of the last chamber to the diameter (d) of the test in two isochronous species of *Valvulineria* (*V. berthelini* Jannin and *V. praestans* Jannin) deriving from the early Cretaceous (Albian) of the Paris Basin. As there are separate areas of dispersion, there is reason to attribute the individuals to two different species of foraminifera. After Jannin (1967, fig. 6)

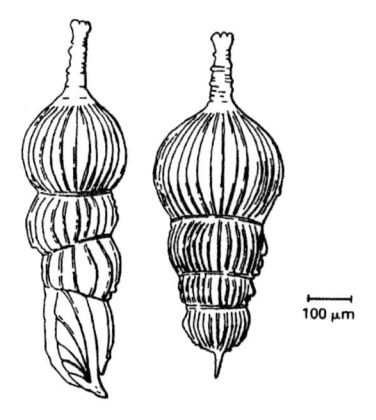

Fig. 13.14 These two Recent foraminifera were originally described with different generic attributions and specific names: on the right, *Nautilus scalaris* Batsch, 1791, subsequently referred to the genus *Nodosaria*; on the left, *Marginulina falx* Jones and Parker, 1860. These individuals, which are always associated, appear to correspond to the microspheric (left) and megalospheric (right) stages of the life cycle of the same species: *Amphicoryna scalaris* (Batsch) (×70). After Buchner (1963, fig. 284)

- By a unimodal curve (Fig. 13.11) when all the specimens are intermediate between the extremes; in this type of continuous variation, the collection is homogeneous and unispecific.
- By a multimodal slope (Fig. 13.12) when the collection is heterogeneous and multispecific.

Dispersion graphs (Fig. 13.13) are also a good method of showing variability, the heterogeneity of the collection being indicated by separated fields.

The species, therefore, is seen, both in biology and palaeontology, to represent a collective, statistical fact, corresponding not to the incarnation of a fixed and arbitrary type but to the morphological variability of a population. Species are delimited by way of discontinuities in this variability although, in certain cases, when it is possible to compare living representatives of a single group, it may be necessary to combine in a single species two or more different forms each relating to the successive stages of a life cycle. Every species, therefore, includes both adults and larvae as far as ostracods are concerned (Fig. 4.7), and in the case of foraminifera (Figs 3.3 to 3.5 and 13.14), the micro- and megalospheric stages.

Species in Time and the Modes of Speciation

What has just been said applies to isochronous populations. In palaeontology, the analysis of populations is complicated by the addition of a further factor: time. With the passage of time, the variability curves for populations are modified, leading to the progressive emergence of new species (speciation).

The term **phyletic lineage** is used to designate a series of species succeeding each other by descent through time. Where it is well established, speciation can be seen to operate through two processes:

- By progressive shift of the variability curves and without increase in the number of species: this is **anagenesis** (Figs 13.15 and 13.17).
- By division of the phylum and multiplication of the number of species: this is **cladogenesis** (Figs 13.16 and 13.18).

Some significant observations have been made on the basis of microfossils. In a continuous succession, any break is arbitrary. Two neighbouring species can be distinguished from each other only with certainty when overlap is absent, at least in the domains of variability of a characteristic. This necessity is not universally recognized and the perennial conflict continues between those specialists who bring things together and those who, whether conscious typologists or not, break them down. Whatever the attitude adopted, the subjective response to the problems raised must be prompted by the objective analysis of variability in populations. It is indispensable to have statistical treatments that cover large numbers of specimens and are as complete as possible. Because of their abundance and normally good state of preservation, microfossils are ideal material for such studies.

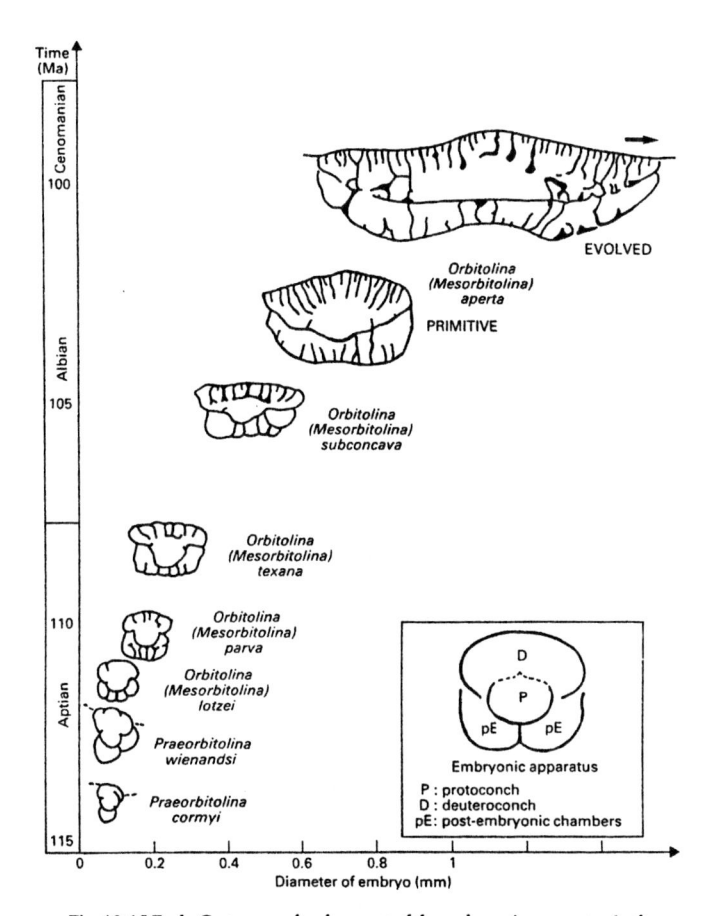

Fig. 13.15 Early Cretaceous development of the embryonic apparatus in the foraminiferal lineage *Praeorbitolina cormyi* Schroeder – *Orbitolina (Mesorbitolina) aperta* (Erman) (×70). This growth accompanied the parallel evolution of:

- The protoconch, which changes from spherical to lenticular.
- The deuteroconch which, initially undivided, divides into primary and secondary chamberlets.
- Post-embryonic chambers, initially undivided and partially enclosing the protoconch, which subsequently divide into chamberlets. After Schroeder (1975, figs 6 and 8)

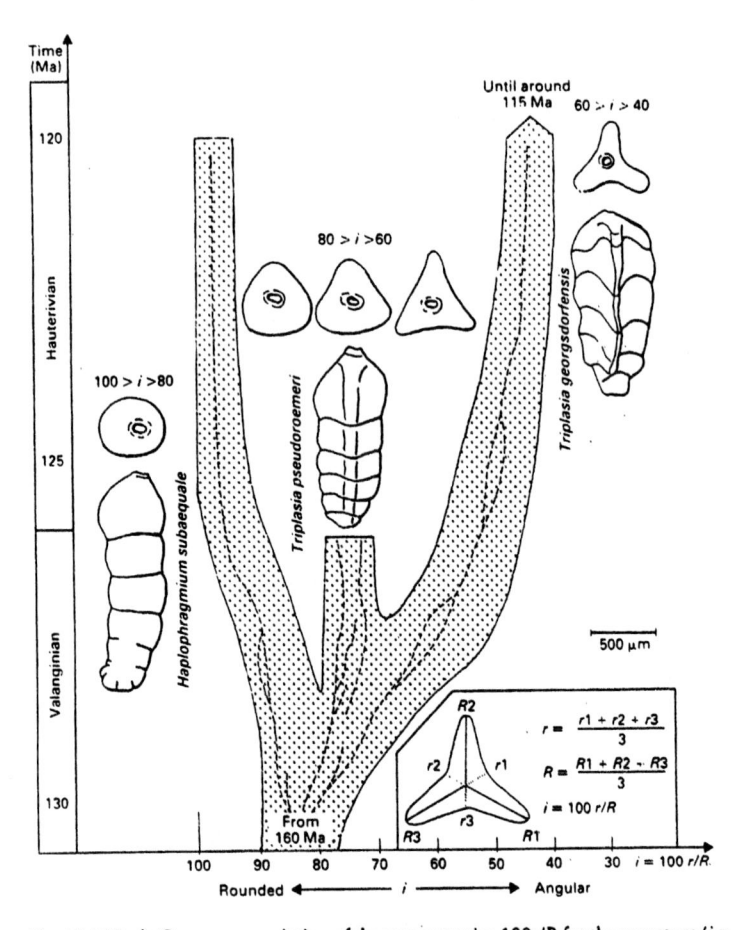

Fig. 13.16 Early Cretaceous variation of the parameter $i = 100r/R$ for the curvature ($i =$ close to 100°) or angularity ($i =$ close to 40°) of the transverse section of the test of agglutinated foraminifera of the genera *Haplophragmium* and *Triplasia*. The subdivision into species suggested by the evolution of the parameter adopted is strengthened by:
- The diminution in the number of chambers.
- The position of the aperture, which becomes terminal.
- The growing curvature of the sutures.

The three species became differentiated around 128 million years (Ma) ago; *T. pseudoroemeri* and *H. subaequale* disappeared respectively around 126 and 120 million years ago; *T. georgsdorfensis* continued until 115 million years ago (×25). After Gerhardt (1963, fig. 22)

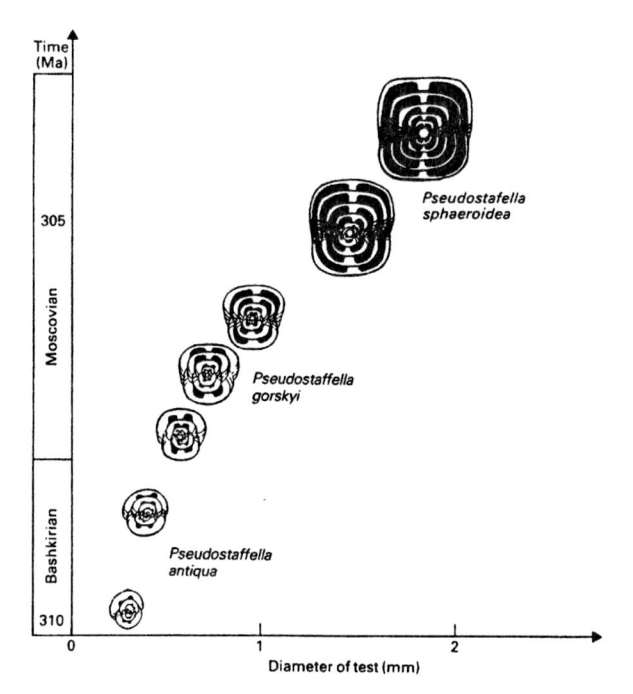

Fig. 13.17 Phylogenesis during the Carboniferous of a fusulinid genus (*Pseudostaffella*) emphasized by the increase in the size of the test as well as by changes in form and growth of the chomata (in black) with time (×35). After Rauzer-Chernousova (1963, fig. 1)

3. EVOLUTIONARY TRENDS

The emergence of species is rarely known with precision. This unfavourable situation is further exacerbated when taxons of a higher order are considered. With foraminifera, for example, most genera and families appear abruptly, without obvious relationship one to another, and in the apparent absence of transition forms.

Fossils embody successful stages of evolution. Although they reveal nothing of the underlying mechanisms, they provide information on the lines of evolution the 'laws' of which have been uncovered through more than a century of observations. These, however, are subject to many exceptions and so must be taken as conveying general trends. Such 'laws' will be illustrated by examples deriving from microfossils.

Wherever they are known, the ancestors of a group are distinguished from later related forms by their small size and simple morphology. This 'law' of non-specialization of early forms is confirmed by the appearance both of the earliest fusulinids and the oldest planktonic foraminifera.

In the course of time, each phylum (evolutionary series) exhibits episodes of accelerated evolution (**tachytelia**) alternating with phases in which evolution is slow or almost at a standstill (**brachytelia**). Since their first appearance, the planktonic foraminifera have been through three periods of diversification – in the late Cretaceous, the Eocene and the Miocene, each one being followed by a relative decline.

During these explosive phases of evolution, and afterwards though more slowly, the phyla become progressively more specialized with:

- A continuous increase in the size of individuals; this is the case with the Fusulinidae (Fig. 13.17) and the genus *Alveolina* (Fig, 13.19) though with *Nummulites* the maximum diameter is reached rapidly only to decline later.
- A growing complexity with the acquisition and development of organs and distinguishing morphological and microstructural features; this is another case where the Fusulinidae and the genus *Alveolina* furnish good examples to confirm the 'law'.

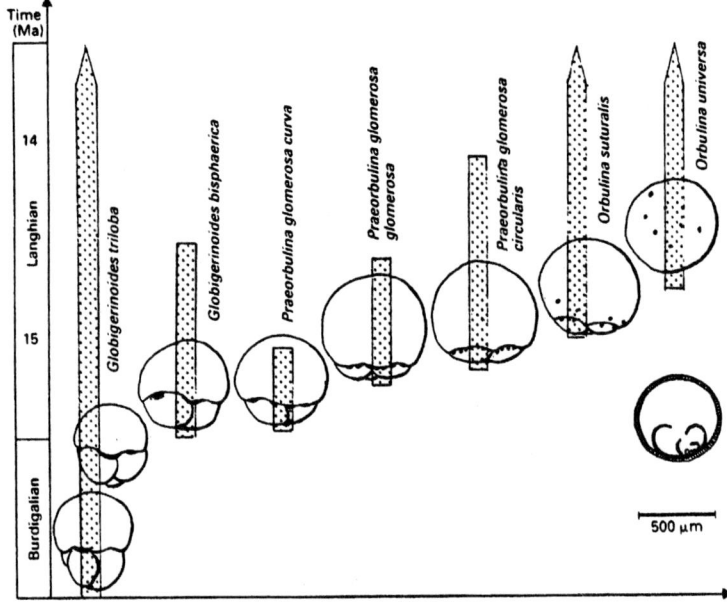

Fig. 13.18 Phylogenesis during the Miocene of a group of planktonic foraminifera evolving from the genus *Globigerinoides* to the genus *Orbulina* through the gradual envelopment of the first chamber by the last. Bottom right, axial section of *Orbulina universa* showing the young stage completely enveloped by the last chamber (×30). After Blow (1956, text-figs 3 and 4)

MICROFOSSILS – THE KEY TO BIOLOGICAL PROBLEMS

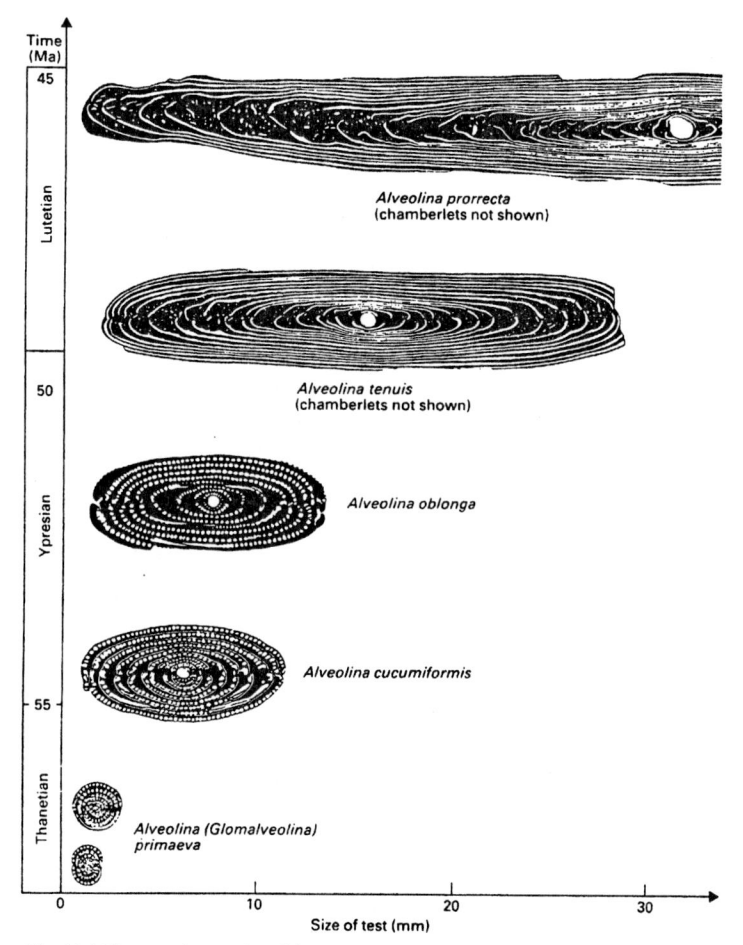

Fig. 13.19 Increase in test size of the genus *Alveolina* (×35) during the Palaeocene and Eocene. This evolution was accompanied by:
- Loss of the initial bunching of the first chambers (at least in megalospheric individuals).
- Increase in volume of the initial chamber.
- Increase in the number of whorls in the spiral.
- Progressive lengthening of the test, leading to a thickening of the floor of the polar sections and consequently to the appearance of chamberlets and secondary floors.
- Increase in the number of chamberlets per chamber.

After Hottinger (1960, pp. 50, 136, 143 and pl. 16)

The progressive specialization of phyla often seems to be oriented (directed, some would say). Orthogenesis such as this is displayed by the *Pseudostaffella* lineage (Fig. 13.17) and by the lineage that led from *Globigerinoides* to *Orbulina* (Fig. 13.18). When the lineages are inextricably interwoven, as for example with the Eocene alveolinids, the overall picture is characterized, nevertheless by fairly clear general tendencies (Fig. 13.19).

Evolution is irreversible and apparent returns may be explained by different rates of transformation for any one of the characteristics under examination. This type of evolutionary 'mosaic' can be seen in species of *Alveolinella* (the modern representatives of the Alveolinidae), which have characteristics that are 'primitive' (continuous intralocular septulae) and others that are 'evolved' (intralocular floors). Associations of characteristics, some tachytelic and others brachytelic, explain why certain apparently 'archaic' taxons persist amidst 'evolved' forms.

The fundamental biogenic 'law' states that the development of the individual (**ontogenesis**) is an abbreviated recapitulation of the evolution of the phyletic lineage (**phylogenesis**) to which it belongs. This is a postulation that has been defended and criticized with equal passion. Although it is true that it cannot always be verified from observation, there are many examples that corroborate it. One such is *Orbulina* (Fig. 13.18) for which both ontogenesis and phylogenesis are known. Most often, however, phylogenesis cannot be established only on the basis of a real succession of intermediate fossil forms. In these cases, it has to be reconstructed from the ontogenetic characteristics of the young (particularly, in the case of foraminifera, for those of the more-conservative form B). In certain fortunate cases, such examples of palingenesis make it possible to see the archetype of a genus or family, as well as the phylogenetic stages that separate the groups (Fig. 13.20).

A final example is the great expansion of calpionellids in the oceans at the end of the Jurassic and in the early Cretaceous, with the concomitant inhibition of the planktonic foraminifera, which did not cease until the former had disappeared. This is a good illustration of the 'law' of relay stages, according to which certain groups are periodically replaced by others that are theoretically more evolved.

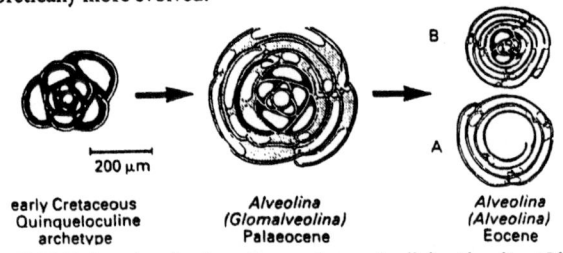

Fig. 13.20 Quinqueloculine juvenile stage known in all the *Alveolina* (*Glomalveolina*) and the B-forms of *Alveolina* (*Alveolina*) suggest that the ancestor of the genus was a miliolid. The transition from this form to that of *Alveolina* probably took place during the Cretaceous with the acquisition of supplementary chambers that were planispiral and divided by septulae. The fact that none of these transitional forms has been found is explained by their small numbers, rapid evolution and localized habitat (×54)

4. MICROFOSSILS AND THE ORIGINS OF LIFE

A Fascinating Problem

Human beings have long been fascinated by the origin of life; for a long time we have had to be content with mythical explanations. The problem was not confronted scientifically until shortly after the Second World War when the biochemists A.I. Oparin and J.B. Haldane elaborated, independently of each other, a biogenic theory that was to be confirmed by S.L. Miller's famous experiment of 1959. According to this hypothesis, biogenesis, which is no longer possible, took place a few thousands of millions of years ago when the surface conditions of our planet were very different from those that prevail today. According to Haldane, there was a 'primeval soup' – an aquatic environment rich in salts covered by an anoxic atmosphere which was transparent to ultraviolet radiation. It was here that the first organic molecules were formed and then concentrated in coacervates. These grew in complexity until, eventually, they formed 'protobiota' and the first living creatures.

Up until 25 years ago, palaeontological discoveries in Precambrian strata were rare. Some of these (e.g. *Eozoon canadense* and *Corycium enigmaticum*) have turned out to be dubious although their significance is a question that continues to be raised from time to time. Others are more firmly established. Karelian shungite, for example, is considered to be a metamorphosed boghead and the stromatolites, which will be discussed later (Chapter 14), were viewed as biosedimentary constructions by their discoverer C.D. Walcott.

This apparent poverty of organisms in the Precambrian resulted from the fact that searchers were looking for large fossils. Once it was recognized that the stromatolites had an algal origin, microfossils began to be sought in thin sections and palynological preparations. The calcareous elements that were examined proved to be sterile but in 1954 S. Tyler and E. Barghoorn discovered numerous carbonate microfossils in the silicified parts of stromatolites. Since then, 45 other deposits, also silicified, have been found, in addition to some in black shales and, in exceptional cases, fine dolomites. The organic origin of these discoveries has been confirmed by chemical and isotopic analysis of the carbonate consituents. The age of the deposits had been established by radiocarbon dating.

Life in the Proterozoic

What is known comes principally from two deposits. The first dates from 850 to 900 million years ago and is situated in the Bitter Springs Formation in central Australia. It has yielded some 50 different microfossils. Some are filamentous in appearance and are reminiscent of bacterial chains, cyanophyte trichomes (Figs 13.21–23) and mycelial hyphae (Fig. 13.24). Others are spherical in shape and occur either in isolation or in masses (Figs 13.25 and 26). They are ornamented with a puzzling 'black spot', which is also known from several microfossils of the Bungle Bungle dolomite (Australia); with an age of some 1500 million years. This 'black spot' has been the subject of lively debate. Some view it as no more than mineral concretion caused by diagenesis, others as a pore, and still others as the trace of an intracellular organelle that

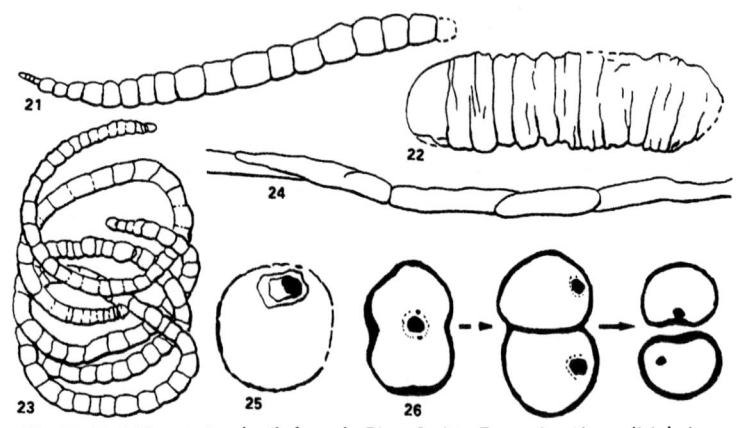

Figs 13.21–26 Some microfossils from the Bitter Springs Formation (Australia) dating from 850 to 900 million years ago

21 *Paleolyngbya barghoorniana* Sch. (×1285)
22 *Caudiculophycus rivularioides* Sch. (×1340)
23 *Anabaenidium johnsonii* Sch. (×1440)
24 *Eomycetopsis robusta* Sch. (×410)
25 *Caryosphaeroides pristina* Sch.: shell with 'black spot' occurring either as an individual or grouped in a mass enveloped in mucilaginous matter (×1440)
26 *Glenobotrydion aenigmatis* Sch.: reconstruction of a sequence reminiscent of mitotic cellular division (×1235)

After Schopf (1968, text-figs 5 and 6, pl. 77, 79, 83 and 85)

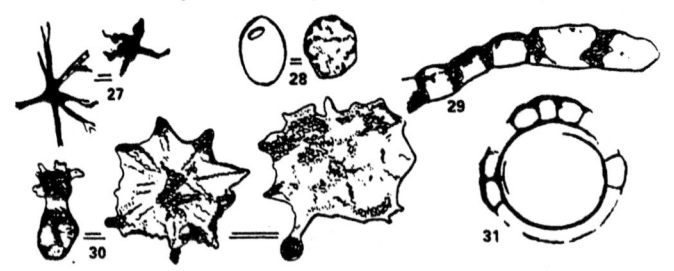

Figs 13.27–31 Some microfossils from the Gunflint Iron Formation (Canada) dating from 2000 million years ago

27 *Eoastrion*: two species with simple or bifurcated appendages (left ×680; right ×885)
28 *Huroniospora*: two individuals, including one (left) with pylome (×2320)
29 *Gunflintia grandis* Barg.: septate filament (×1240)
30 *Kakabekia umbellata* Barg.: three different individuals showing the bulb, stipe and umbellar crown (×1240)
31 *Eosphaera tyleri* Barg.: two concentric spheres linked by trabeculae (×960)

After Barghoorn & Tyler (1965, figs 4, 6, 7 and 8)

might be a pyrenoid or a nucleus. The latter interpretation, which is defended by J.W. Schopf, implies that the spheres are the remains of eukaryotic cells. This opinion is contested by E. Boureau whose examinations of deposits at Richâts in Mauretania with an age of 800 million years make it possible to affirm that, at this date, only prokaryotes were in existence. These were in the process of acquiring through intracellular incorporation the symbiotic cells that would subsequently develop into mitochondria and chloroplasts.

Whatever the case may be, metazoans are known with certainty at the end of the Precambrian (650 ± 50 million years) from the fauna of Ediacara (Australia). What is lacking, however, is physical evidence, from the period between this deposit and those of Bitter Springs and Richâts, to explain the increase in cellular complexity, the appearance of predator protists and the emergence of multicellular organisms.

The second deposit, the Gunflint Iron Formation which outcrops to the north of Lake Superior (Ontario, Canada), is older than the first, dating from 1900 ± 200 million years. It has yielded ovoid shells with (Fig. 13.28) and without pylome, septate (Fig. 13.29) and non-septate filaments, and star-shaped bodies (Fig. 13.27) whose appearance is reminiscent of colonies of ferro-oxidizing bacteria of the genus *Metallogenium*. To these must be added more enigmatic forms, the *Eosphaera* (Fig. 13.31) and the curious *Kakabekia* (Fig. 13.30) living representatives of which have been found in ammoniacal environments.

Life or Protolife in the Archaean?

As far as the period before the Proterozoic is concerned, Archaean strata contain a few occasional stromatolites. Although these have yielded no microfossils, they prove that life had already appeared between 2500 and 3000 million years ago and, despite the controversy surrounding them, they may confirm two recent discoveries.

Siliceous and schistose beds intercalated between the basalts and pyroclastites of the Bulawayan System of Swaziland have produced tiny hollow spheres with a carbonaceous wall. Because of the absence of structure and the variability in size (average: 18 μm; extremes: 1 to 203 μm) of these *Archaeosphaeroides* (Fig. 13.32), it is difficult to assess their significance. It is possible, however, that they are the authentic remains of very primitive living things.

Figs 13.32, 33 The oldest known fossils
32 *Archaeosphaeroides barbertonensis* Sch. & Barg.: Fig Tree Series of the Bulawayan System (Swaziland); age, 3000 million years (×745)
33 *Isuasphaera isua* Pflug: filament (left), branching colony (?) (right); Isua Iron Formation (Greenland); age, 3800 million years (×615)

After Pflug (1976, table 2, fig. 5) and Pflug & Jaeschke-Boyer (1979, fig. 1)

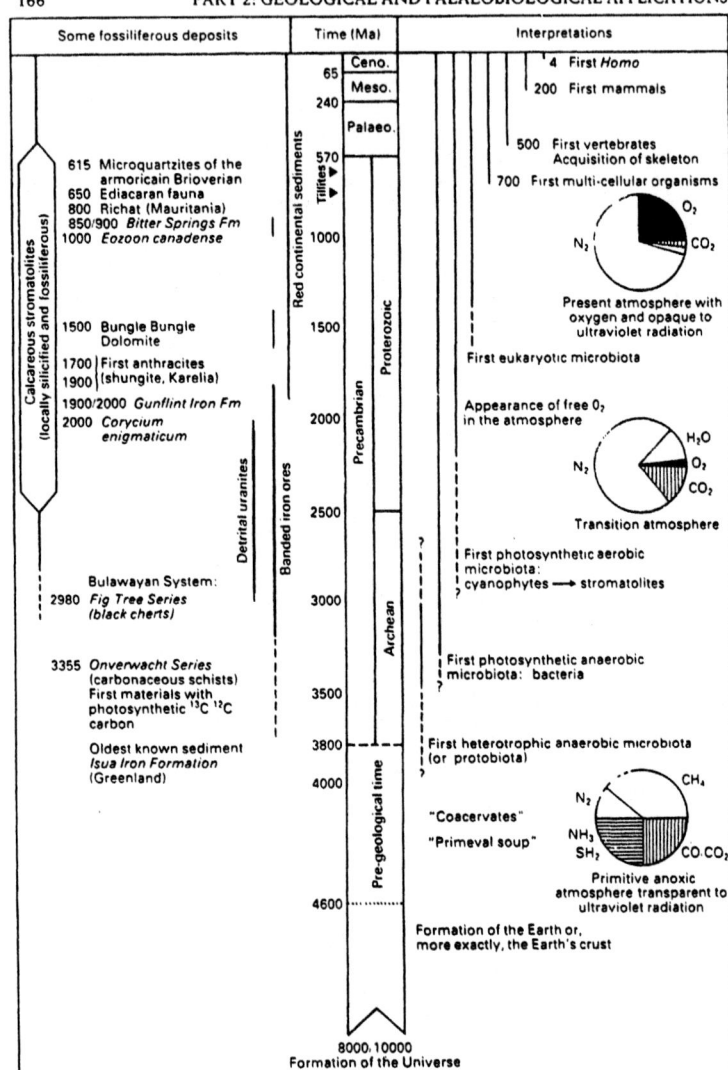

166   PART 2: GEOLOGICAL AND PALAEOBIOLOGICAL APPLICATIONS

Some fossiliferous deposits	Time (Ma)		Interpretations
	65	Ceno.	4 First *Homo*
	240	Meso.	200 First mammals
	570	Palaeo.	500 First vertebrates — Acquisition of skeleton
615 Microquartzites of the armoricain Brioverian			700 First multi-cellular organisms
650 Ediacaran fauna			
800 Richat (Mauritania)			
850/900 *Bitter Springs Fm*			
1000 *Eozoon canadense*	1000	Proterozoic (Red continental sediments)	Present atmosphere with oxygen and opaque to ultraviolet radiation (O_2, N_2, CO_2)
1500 Bungle Bungle Dolomite	1500		First eukaryotic microbiota
1700 First anthracites (shungite, Karelia)			
1900/2000 Gunflint Iron Fm			Appearance of free O_2 in the atmosphere
2000 *Corycium enigmaticum*	2000	Precambrian	Transition atmosphere (H_2O, N_2, O_2, CO_2)
	2500		
Bulawayan System: 2980 *Fig Tree Series* (black cherts)	3000	Archean	First photosynthetic aerobic microbiota: cyanophytes → stromatolites
3355 *Onwerwacht Series* (carbonaceous schists) First materials with photosynthetic $^{13}C\ ^{12}C$ carbon	3500		First photosynthetic anaerobic microbiota: bacteria
Oldest known sediment *Isua Iron Formation* (Greenland)	3800		First heterotrophic anaerobic microbiota (or protobiota)
	4000	Pre-geological time	"Coacervates" "Primeval soup" (N_2, NH_3, SH_2, CH_4, $CO\cdot CO_2$) Primitive anoxic atmosphere transparent to ultraviolet radiation
	4600		Formation of the Earth or, more exactly, the Earth's crust
	8000, 10000		Formation of the Universe

Calcareous stromatolites (locally silicified and fossiliferous)
Detrital uranites
Banded iron ores
Tillites

Fig. 13.34 Stratigraphy of the main fossiliferous deposits of the Precambrian and a possible interpretation of the palaeontological data. It should be noted that the chronology of the principal stages of the evolution of the first living things is highly controversial. E. Boureau, for example, places the appearance of eukaryotic organisms later (between 800 and 700 million years) than it is indicated here.

Finally, microfossils with carbonaceous walls (Fig. 13.33) have been found in the oldest sediments that are known – the cherty quartzites of the Isua Iron Formation in Greenland. They are of a size and form that is to some extent reminiscent of modern yeasts.

The Precambrian, Era of Microscopic Life

It is hardly surprising that the first manifestations of life should be represented by creatures whose affinities remain obscure. Some idea of their mode of existence can be obtained from the analysis of sediments laid down at the time on the continents or in shallow waters. Before 2000 million years ago, they were emplaced in an anoxic environment and include sedimentary uranites and banded iron ores. After this period, the presence of red continental sediments indicates the appearance of free oxygen in the atmosphere.

It is clear then that Archaean microbiota must have had an anaerobic existence. The oldest of them (protobiota?) are thought to have fed off the 'primeval soup'. Once this was exhausted, the still anaerobic microbiota would have had to acquire an autotrophic metabolism, analogous to that of modern bacteria living today. The production of oxygen began with the development of the first photosynthetic microbiota (cyanophytes) of which traces are to be found in stromatolites (Fig. 13.34).

Microfossils and the Possibilities of Extraterrestrial Life

The human imagination is excited by the possibility of extraterrestrial life. Up to the present time, however, no conclusive evidence for it has been put forward. The samples of lunar rock that have been brought back on spacecraft have yielded no trace of life. As for the evidence of meteorites, carbon molecules, some of which are analogous to those of organisms presently living on the surface of our planet, have been found in a very few 'stony' meteorites (only five cases are known!). Are these constituents abiotic, protobiotic or biotic? To answer this question, the meteorites were investigated for microfossils. In 1961, G. Claus and B. Nagy announced that they had discovered numerous small 'organized bodies' in two stony meteorites, one of which had fallen at Orgueil in the south-west of France in 1864.

For the most part, these microfossils resulted from terrestrial contamination but nine of them, ranging in size from 4 to 30 μm, were completely different from any creature or fossil known on our planet. They could, therefore, be 'microfossils indigenous to the meteorites' and so, according to their discoverers, the remains of extraterrestrial organisms. This conclusion has not found acceptance as specialists have regarded Claus and Nagy's material as being, in its entirety, no more than artefacts or contamination from pollens and terrestrial microorganisms.

CONCLUSION

The ecological study of fossil microorganisms is not of interest merely to palaeobiologists. It will be shown later that geologists can also benefit in the interpretation of ancient sedimentary environments and in palaeogeography.

Microfossils also help to illustrate the concept of species and the modalities of evolution. In this context, they provide clues that can be applied to more complex groups of organisms. Furthermore, the existence of some of them, such as the holuthuroids, the ascidians and the conodontophores is attested only by microscopic remains.

Photosynthetic microorganisms have always been, and remain, at the base of the food chain and, as such, are indispensable to the survival and development of all living creatures. Despite all the uncertainties and difficulties that are involved, the study of microfossils as the first manifestations of life is a singularly attractive enterprise since it provides a glimpse of our oldest ancestors and an indication of the vital part that they played in preparing the framework within which their descendants would subsequently evolve.

BIBLIOGRAPHY

D.V. Ager, *Principles of Palaoecology* (McGraw-Hill, New York, 1963), R.T. Hecker, *Bases de Paléoécologie* (translation of a Russian work published in 1957; Technip, Paris, 1960), C. Babin, in *Eléments de Paléontologie*, pp. 64–158, Colin, Paris, 1971), J. Rojer, in *Paléontologie generale*, pp. 104–159, (Masson, Paris, 1974), and also J.W. Valentine, *Evolutionary Palaeoecology of the Marine Biosphere* (Prentice-Hall, New Jersey, 1973). All give a good overview of palaeoecology though without frequent mention of microfossils. Scattered references to the ecology and palaeoecology of microbiota are to be found in works cited in the section on systematic classification in this book. Also worth a mention is J.W. Murray *Distribution and Ecology of Living Benthic Foraminiferids* (Heinemann, London, 1973).

Standard works on biology, zoology and botany usually devote a few paragraphs to the notion of species, speciation and the modalities of evolution. These may be complemented by S.J. Gould *Ontogeny and Phylogeny* (Belknap, Harvard UP, 1977).

The application of statistical methods to microfossils has been the subject of several publications including M. Caron, *Eclogae Geologicae Helvetiae* 60 (1) 47–49 (1976), and F. Magniez-Jannin, *Reviews of Micropaleontology* 10 (3), 153–178 (1967) and *Cahiers de Paléontologie* (1975).

Documentation on the evolution of microfossils is scattered through many stratigraphical works. Though already old, it is worth consulting G.H.R. Von Koenigswald *et al.*, *Evolutionary Trends in Foraminifera* (Elsevier, Amsterdam, 1963), and also R.A. Reyment, *Introduction to Quantitative Ecology* (Elsevier, Amsterdam, 1971).

Finally, M.G. Rutten, *The Origin of Life by Natural Causes* (Elsevier, Amsterdam, 1971) is a good synthesis, a large part of which concerns microfossils. This can be supplemented by more recent works such as F.T. Banner and F.M. Lowry (Eds), in: Evolutionary case histories from the fossil record, *Special Papers in Palaeontology* (Palaeontological Association, in press), J. Brooks & G. Shaw, *Origin and Development of Living Systems* (Academic Press, London, 1973) and C. Ponnamperuma, *Chemical Evolution of the Early Precambrian* (Academic Press, London, 1977).

Chapter 14

Microfossils as a Source of Sediments

Although macrobiota, in particular corals and calcareous algae, play a considerable part in the formation of rocks, microbiota are responsible for much thicker and more varied sedimentary formation. The remains of both types accumulate in sediments but, in addition to this direct contribution, living forms participate in lithogenesis indirectly in ways that are still little appreciated; this is especially true of microbiota.

1. LITHOGENESIS THROUGH BIOCLASTIC ACCUMULATION

By **bioclast** is meant any biogenic remains entering into the constitution of sedimentary rocks, whether they are mineralized or not, and whether they remain whole or in fragments.

Bioclastic Sedimentation

In recent environments, there are many examples of bioclastic sedimentation. The largest by far is the deposition on the ocean floors where the remains of planktonic microorganisms fall in a continuous rain from the surface waters. If the floor is above the CCD, calcitic skeletons are not corroded and they accumulate to form an **ooze** of coccoliths and planktonic foraminifera. Below the CCD, the calcitic skeletons disappear and there is a concentration of diatom frustules and radiolarian tests. Calcareous and siliceous oozes cover two thirds of the ocean bed (Figs 14.1 and 2). A single milligram of siliceous ooze contains thousands of radiolarian tests or hundreds of thousands of diatom frustules. Every 1000 years, some 10 kg of siliceous ooze is deposited per square metre of ocean surface. Calcareous sedimentation reaches 20 kg/m^2

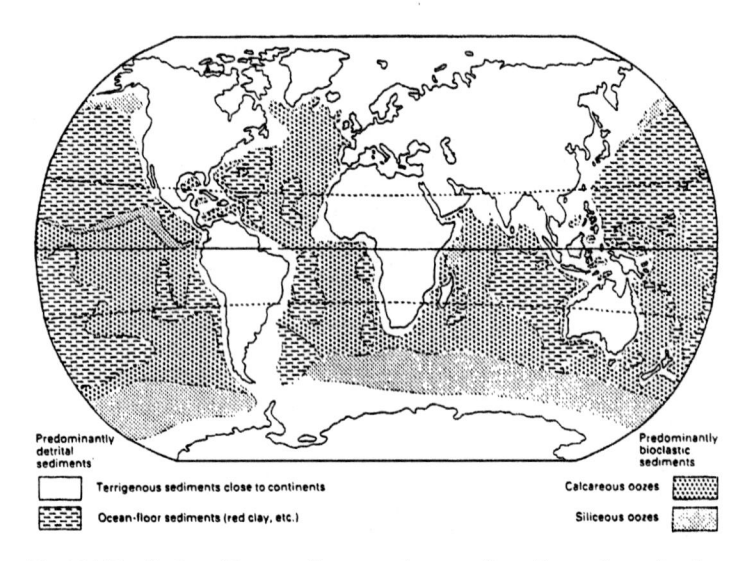

Fig. 14.1 Distribution of Recent sediments on the ocean-floor. Pteropods are abundant only in the calcareous oozes of the South Atlantic. The siliceous oozes of the North Pacific and the Antarctic consist predominantly of diatoms. Those of the tropical Pacific are rich in radiolarians. Largely after Arrhenius (1963, fig. 1)

Type of sediment		Bioclastic content	Chemical composition (%)		Ocean surface covered		Average depth (m)	Sedimentation (mm/1000 years)
			$CaCO_3$	SiO_2	$10^6 km^2$	%		
Calcareous oozes	Pteropod	< 30% pteropods	74	2	1.5	0.5	2070	5–20
	Coccolith + foraminiferal	< 50% coccoliths > 30% foraminifera	80	2	128	47.2	3610	
Siliceous oozes	Diatomaceous	> 50% siliceous tests	4	47	31	11.6	3900	4–5
	Radiolarian		2.5	70	7	2.6	5290	

Fig. 14.2 Composition and distribution of Recent bioclastic sediments on the ocean floor.

at a depth of 2500 m. The average rate of sedimentation in the Pacific is estimated, for a millennium, at 4 to 5 mm of siliceous ooze and 5 to 20 mm of calcareous ooze. The subsequent compaction is in the order of 20 to 40% of the initial volume.

In certain places, the neritic domain is also the recipient of considerable bioclastic sedimentation. The sands that surround coral reefs are composed of the tests of foraminifera (8 to 10%), small fragments of mollusc shells (10%), corals (25 to 30%), thalluses of corallinaceous red algae (25 to 30%) and crystals of aragonite deriving from the breakdown of the thalluses of *Halimeda* and other udoteacean algae (18 to 30%).

In lakes, the phenomenon is much more modest. In some cases, the accumulation of diatom frustules forms a deposit with an appearance and consistency that have led to its being called 'organic felt'. In other lakes, colonies of *Botryococcus* accumulate to form slimes or **sapropels** that are rich in organic matter and which evolve, after dehydration, into saprocols of a rubbery consistency.

Ancient Bioclastic Sediments

Accumulations such as those described above are found in ancient sedimentary series. The most significant example is provided by the **diatomites** (formerly called kieselguhrs or tripolis). These loose siliceous rocks are exclusively formed from diatom frustules, both whole and broken, pressed against each other. The thickness of the rocks can be considerable and may, in places, reach several hundreds of metres.

Diatomites are light rocks with a density when dry of 0.5. They are soft and, though coherent, also friable. They owe their remarkable permeability and porosity (up to 80%) to the fineness of the frustules and the spaces between them in the absence of cement. As they are capable of retaining impurities (e.g. bacteria and colloids) of less than 0.2 μm, diatomites are used industrially for the filtration of drinks, medicines, waste water, etc. For this purpose, some 2 million tons are extracted annually. The main suppliers are the USA, the USSR and France (Pleistocene deposits at la Bade (Cantal), Pliocene at Foufouilloux (Cantal), and Miocene at Saint-Bauzile (Ardèche)).

Many fine-grained limestones, both loose and compact, which appear, under the light microscope, to be homogeneous can be seen, in the SEM, to be formed from an accumulation of microfossils. These biogenic micrites are called **nanoagorites** (from Greek *nano* and *agora* = assembly). The 'ammonitico rosso' and 'majolica' of the Italian Mesozoic are almost entirely formed from *Nannoconus*. The chalks of the Upper Cretaceous in the Paris Basin are not true nanoagorites because up to 50% of their volume may be formed from coccoliths, *Nannoconus* and pithonellids. The bonding phase includes monocrystalline calcitic particles. These are very small (2.5 to 0.1 μm) and are without sharp edges. Their origin is the subject of debate (Fig. 14.3).

Many other rocks not entirely formed from microfossils contain, nevertheless, a considerable proportion of them in their cement which varies in its composition:

- Micritic or sparitic in various limestones (biomicrites, biosparites, lumachels (fire-marbles) containing large foraminifera.

- Argillaceous in microfossiliferous marls.
- Siliceous in radiolarites (jaspers and lydians) and in spongoliths (gaizes).
- Phosphatic in certain phosphorites.
- Carbonaceous and amorphous in coals of every origin, sapropelic (cannel coals, bogheads) or lignocellulosic (certain coals).

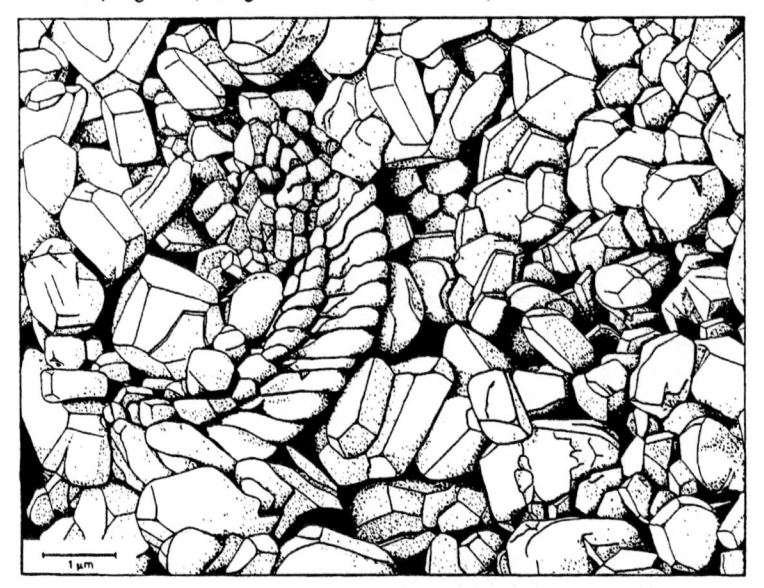

Fig. 14.3 Nanofacies of chalk on a fracture surface. The coccolith that can be seen in the left half of the figure is embedded in a matrix of very small calcitic particles with 'spaces' between them. Chalk from the Upper Cretaceous (Senonian) of the Paris Basin Drawing after an original photograph.

2. LITHOGENESIS THROUGH THE CONCENTRATION OF AMORPHOUS ORGANIC SUBSTANCES

Concentration of Mineralized Substances

The siliceous cement of radiolarites and spongoliths may be formed – at least in part – from silica released by the dissolution of the radiolarian tests and sponge spicules on the sea-bed or within the sediment during diagenesis.

This hypothesis seems to be applicable to other rocks. It is impossible not to be surprised by the contrast between the enormous abundance of crystals of aragonite deriving from algae in modern tropical sediments, and their total absence in ancient peri-reef limestones. The metastable aragonite must have dissolved and then reprecipitated in the form of a stable calcitic cement. The same logic may explain why phosphate deposits are so rich in dinocysts. Dinoflagellates would have played a similar lithogenic part by concentrating phosphorus in their cells and later releasing it on the sea-bed as their protoplasm decomposed.

It can bee seen, therefore, that the contribution of organisms to lithogenesis is not confined merely to the supply of bioclasts. Organisms, and particularly microbiota, are known to select and concentrate chemical elements (e.g. Ca, Si, Mg, P and Fe), which are more or less dissolved in the surrounding environment. When they die, these elements are released and may be incorporated in sediments.

Concentration of Organic Substances: Carbonaceous Rocks

There is no longer any doubt that all carbonaceous rocks have a biotic origin. Petrographic analysis of coal (using thin sections or more often polished surfaces observed in reflected light under a metallographic immersion microscope) shows spores, remains of leaf cuticles, fragments of secreting tissue and woody fibres, walls of sclerenchymatous cells, fungal sclerotia, etc. These microfossils are embedded in a groundmass of vitrinite, a homogeneous amorphous substance that appears grey in reflected light and red in transmitted light (Fig. 14.4). The microfossils show that the coals had their origin in the accumulation of terrestrial plant debris in an anoxic lacustrine environment. The vitrinite itself must have resulted from the flocculation of colloidal solutions deriving from the incomplete decomposition of cytoplasm and cellulose walls.

Total world coal deposits are estimated to amount to 5×10^{12} tonnes. Although this figure is spectacular, it represents no more than a thousandth of the organic constituents contained in sedimentary rocks. Palynological analysis shows these to be in the form of kerogen – palynomorphs and amorphous substances more or less finely dispersed. The organic constituents that predominate in sapropelic coals form a not negligible part of dark bituminous rocks, whether these are calcareous (e.g. banded limestone of the Kimmerid-

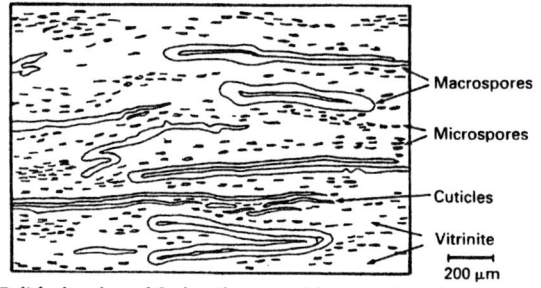

Fig. 14.4 Polished surface of Carboniferous coal from northern France, perpendicular to the stratification and seen in reflected light. After Duparque (1933, pl. 8)

gian at Orbagnoux (Ain)) or argillaceous (e.g. black shales Palaeozoic ampelites, 'schistes-cartons' of the Toarcian in the Paris Basin). The amorphous part of the kerogen comes from protoplasmic constituents that have not been completely broken down. Phytoplankton, especially diatoms and dinoflagellates, are the main primary producers of organic matter but it is important not to underestimate the contribution of bacteria, zooplankton, benthic algae and fish. Diatoms produce sudden and extraordinary 'blooms', seen in the 'mare sporco' of the Adriatic and dinoflagellates produce 'red tides' in the sea and the 'river of blood' in the Nile with densities exceeding several millions of individuals per litre of water. These phenomena play a significant part in the concentration of organic matter in sediments. The multitudes of microbiota consume the oxygen in the water rendering it toxic to molluscs, fish and birds. As these die, vast quantities of organic matter are deposited in sedimentary environments which are, in consequence, rendered anoxic.

The proportion of organic matter that is preserved and incorporated into sediments is low and does not rise above 1%. Although this is only a tiny fraction, it is of vital importance for humanity, being the source of our fossil fuels; to a large extent these owe their formation to the action of microbiota.

3. THE LINK BETWEEN LITHOGENESIS AND MICROBIOTIC ACTIVITY

Organisms also play a part in the construction of rocks through more direct activity. Encrusting algae contribute actively to building coral reefs. Apart from exceptional cases, the participation of fixed foraminifera is very slight. Nevertheless, in the Eocene series, beds of *Nummulites* lying parallel to the palaeocoastlines are suggestive of reefs (see Fig. 16.1)

Stromatolites are laminated sedimentary structures which vary in area from a few millimetres to 80 m with a thickness of 20 m. when fixed, they are stratiform or columnar (Fig. 14.5), while free forms are spheroid (oncolites). In thin section, it can be seen that some of these structures consist of bundles of fine tubes, representing the calcified sheaths of filamentous blue-green algae (cyanophytes). Most, however, are without this feature and are termed cryptalgal. Whatever the case may be, carpets or mats of filamentous cyanophytes are responsible for the periodic growth of stromatolites (Fig. 14.6) by trapping fine detrital particles and by the precipitation of carbonates as a result of photosynthesis. The construction is thus a biosedimentary one within which the cyanophytes are only preserved when their sheath is calcified.

Stromatolites were abundant in the Precambrian; they are evidence of the extremely early (around 3000 million years) appearance of cyanophytes as well as associations of various microfossils. They continue to be frequent in the Palaeozoic and the Jurassic when they colonized neritic marine environments. The development of more evolved algae, especially the rhodophytes, gradually relegated the cyanophytes to less favourable habitats – intertidal and lacustrine. Only 20% of Recent cyanophytes have a marine habitat. Although they still construct stromatolites, they seem to have lost the capacity to calcify their sheath, an ability maintained only in a few lacustrine cyanophytes – *Schizothrix, Scytonema* and *Rivularia*.

MICROFOSSILS AS A SOURCE OF SEDIMENTS 175

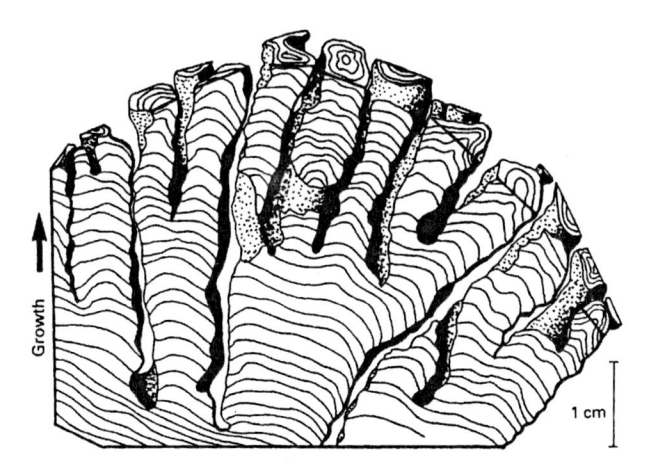

Fig. 14.5 Reconstruction of a columnar (or colloform) stromatolite from the Precambrian (Bitter Springs Formation) of Australia (×4/3). After Walter (1972, fig. 42)

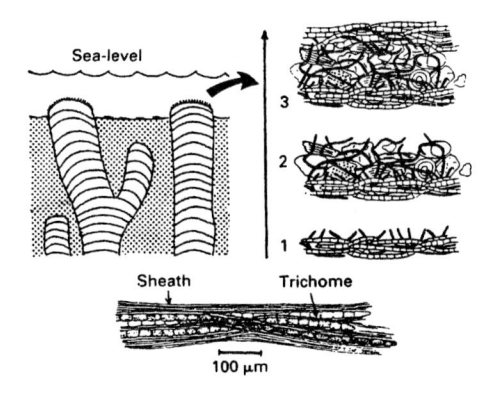

Fig. 14.6 Discontinuous laminar growth of a living stromatolite by the alternation, at the ends of the columns, of development phases of algal filaments (acting as filters) and phases trapping small detrital particles. This structure is also the seat of precipitation of carbonates produced by photosynthesis. Below, details of living algal filaments (*Schizothrix*) with a non-calcified sheath (× 100). After Gebelein (1969, fig. 14) and Bourrelly (1970, fig. 131.6)

Bacteria (and/or Cyanophytes) which currently participate actively in linking the phases of geochemical cycles, have played a similar role in the past. They modified the physicochemical characteristics of sedimentary environments, and permitted the accumulation and preservation of the mineral constituents produced by their metabolism as well as the organic substances deriving from the breakdown of their protoplasm. Their lithogenic role is manifest in various types of rocks and sedimentary ores. Some of these yield tiny bodies or bacteriomorphs which are suggestive of fossil bacteria but about which it is difficult to come to firm conclusions. They are, in fact, difficult to determine because of their size (a few micrometres), their morphological simplicity and their present abundance. Some bacteriomorphs may be no more than contaminations by modern bacteria which are either in suspended animation or dead and mummified. This is a matter for research by specialists in microbiology, geochemistry and micropalaeontology, as the fossil bacteria appear to reveal themselves as veils or films of organic matter.

CONCLUSION

Biolithogenic processes have been treated separately simply for didactic purposes. In fact, in any given sediment, they interfere with each other and with other abiotic processes. Following G. Deflandre (1967), we would emphasize the fundamental importance of the role of the microbiosphere in view of 'the disproportion between the grand scale of the results and the tiny initial causes, between the volume of rocks engendered and the minimal size of the creatures generating them'.

BIBLIOGRAPHY

Manuals of sedimentary petrography vary in the space that they devote to the lithogenic activity of microbiota. Apart from older works by L. Cayeux, reference is made to A.V. Carozzi, *Microscopic Sedimentary Petrography* (Wiley, New York, 1960); A.S. Horowitz & P.E. Potter, *Introductory Petrography of Fossils* (Springer, Berlin, 1971) and also G. Deflandre, *La Vie Créatrice de Roches, Que sais-je?* (PUF, Paris, 1967) which, although out of date, makes good reading.

On stromatolites M.R. Walter (ed.), *Stromatolites* (Elsevier, Amsterdam, 1976) and C. Monty (ed.) *Phanerozoic Stromatolites* (Springer, Berlin, 1981) give an overview of these biosedimentary structures.

The geological importance of microbes is shown by J. Zajic, *Microbial Biogeochemistry* (Academic Press, New York, 1969). The article by M. C. Janin and G. Bignot, Microfossiles thallophy tiques des concrétions polymétalliques laminées, *Revue Micropalaeontogique* 25 (4) 251–264 (1983) is an example of current research on fossil bacteria.

Chapter 15
Microfossils − Chronometers of the Phanerozoic

During the eighteenth century, the idea grew that every sedimentary layer or stratum could be characterized by the fossils that it contained. It was not long before the first biostratigraphers (A. Brongniart, 1823) were directing attention to the nummulitid foraminifera. E. Renevier used the term nummulitic to designate one period of the Cenozoic but this has now been abandoned in favour of the name Palaeogene.

In the period before 1940, stratigraphers still relied entirely on foraminifera (especially the larger ones), some ostracods, and, for the sediments of the Pleistocene, pollens. Although the results they obtained were promising, they were not always received without scepticism. The rapid development of microbiostratigraphy, which followed the end of the Second World War, stemmed mainly from:

- the definition of **microfacies** by J. Cuvillier who commenced a methodical examination of them in 1945; and
- the use of **biozonation**, an idea outlined by N.N. Subbotina and then generalized from 1957 onwards by H. Bolli and other research workers.

1. MICROFACIES

In defining the microfacies as the set of mineralogical and palaeontological characteristics contained in a sedimentary rock as seen under the light microscope at magnifications in the order of a few dozen times, J. Cuvillier was going back to the classical conception of facies and at the same time enriching it. A microfacies is examined several times before it is defined. The first examination deals with the rock matrix and the mineral elements, whatever their origin might be. The second consists of the determination of the bioclasts. These examinations are then complemented by observation of the relative frequency of the constituents, their size, mutual arrangement, etc.

The use of microfacies for the purpose of stratigraphy aroused some concern. It is, of course, well known that a facies is not characteristic of a particular age and that it may be repeated in apparently identical form throughout geological time. Would not the same hold good for microfacies?

Research conducted by Cuvillier showed, however, that microfacies are in fact much more diversified than facies, the characteristics taken into consideration being both more numerous and more conclusive. The probability of two different formations having identical microfossils is virtually zero. So powerful is the method that it can be used not only to date but also to determine the geographical provenance of rocks. For this reason, it is commonly used to identify the materials used as tools in prehistoric times, and used in the construction of the monuments of antiquity and the Middle Ages.

2. BIOZONES AND BIOZONATIONS

Significance of Biozones

Instead of considering – as with the microfacies – the palaeontological assemblage in its entirety, it is more convenient to focus attention on the representatives of a few groups that are known to have a wide geographical distribution and a sufficiently rapid evolution.

These criteria are satisfied by the microfossils, particularly planktonic forms, and accordingly they can be used to divide time into **biozones**. These are of several types (Fig. 15.1):

- **Acrozone** (from the Greek *akros* = culmination) or **total range zone** is used to designate the total extension of a taxon. The terms *acme* or *epibole* are used, though less frequently, to signify that part of the acrozone in which the numerical development of the taxon is at its maximum.
- **Concurrent range zone** indicates the simultaneous presence of two or more taxa; **cenozone** (from the Greek *koinos* = common) is used when, as in the case of the microfacies, the totality of the microfossils is being considered.
- **Interval zone** refers to a zone bounded by two events and coinciding neither with coexistence nor extension of taxa.

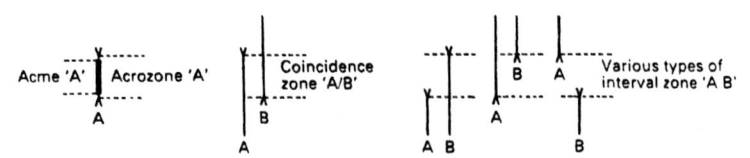

Fig. 15.1 Different zones used in biostratigraphy. Acrozones and concurrent zones are based on criteria of presence. Interval zones are, at least in part, based on criteria of absence and are bounded by:

- disappearances of A and B;
- appearances of A and B; and
- disappearance of B and appearance of A

Every zone bears the name of its index taxon and this is frequently preceded by a symbol (e.g. P for Palaeogene foraminifera, N for Neogene foraminifera, NP for nanoplankton, W for *Wetzeliella*).

Biozonations

In practice, the first step is to establish empirically the distribution of the representatives of the chosen group in the strata of a given region. Once the results are in the form of a graph (Fig. 15.2), the series is subdivided on the basis of the palaeological events (appearance and disappearance of taxa) that emerge.

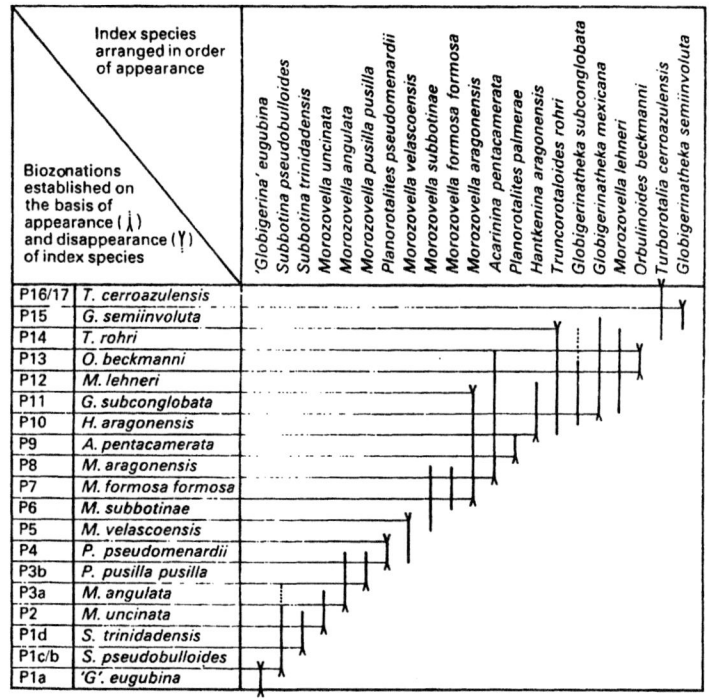

Fig. 15.2 Succession of some planktonic foraminifera of the Palaeocene and the Eocene arranged in order of appearance. These index species (or taxon guides), whether by appearance or disappearance, define reference events that serve to divide time up into biozones. Some biozones (e.g. P1a, P4) are acrozones. Others are interval zones bounded either by disappearances (e.g. P5), by appearances (e.g. P1b/c) or by a disappearance and an appearance (P12). This zonation, which is valid with slight modifications for the whole Mesogean Domain, was established by H. Bolli in 1957 and was completed in 1966 from isolated outcrops and drillings on the island of Trinidad (West Indies)

As soon as the first investigations began, many zones, particularly those relating to planktonic microfossils, were found at widely separated geographical points. Once partial sections were collated, it was easy to list uninterrupted successions of zones or biozonations that were valid over vast areas. The most exact and the most widely used of these are based on planktonic foraminifera (80 zones from the early Cretaceous to the present), coccoliths (90 zones from the Lias to the present), conodonts (140 zones from the Ordovicians to the end of the Triassic) and chitinozoans (40 zones in the Ordovician and the Silurian).

In the beginning, zonations were established independently of each other but then the parallels between them were revealed through the analysis of deposits containing representatives of at least two groups of significant microfossils. The correlations that were obtained showed where the boundaries between zones of different successions failed to coincide.

Limits and Value of Biozonations

If the zones are to have any chronological value, the biological events by which they are defined must be synchronous wherever they are detected. Unfortunately, it cannot be stated without reservation that this is the case.

The first problem is that the establishment of zones requires clearly defined species that are easily and uniformly determined no matter who is responsible for the determination. However, in the hope of making the method more precise, the guide or index taxa have been multiplied with the result that determinations have become difficult and arbitrary, authors being unable to agree with each other about them or about the events that can be deduced from them.

A further point is that the events in question are not simply the result of biological evolution. Many are, in fact, ecological in character and depend on the evolution of conditions in the environment. It is possible for the

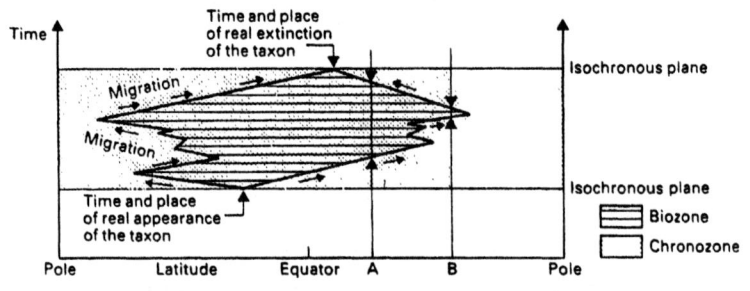

Fig. 15.3 Relation between a biozone (in this case an acrozone, a biostratigraphic unit defined by the extension in time and space of a taxon) and a chronozone (a chronostratigraphic unit independent of biological phenomena and valid world-wide). The geographical extent of the biozone varies with time and, more seriously, duration also varies with latitude. At points A and B, the duration of the biozone is different as it is bounded by the moment of arrival and the moment of departure of the taxon at these two points as a function of variations in the environment

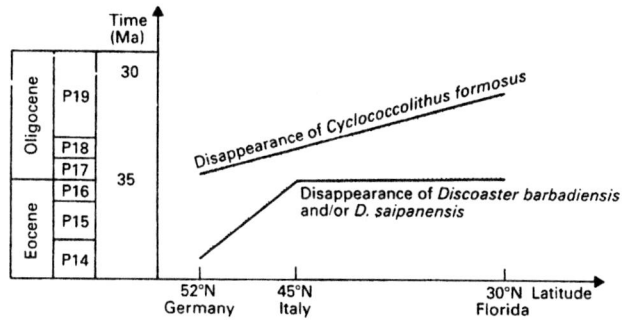

Fig. 15.4 Position and duration of the coccolith biozone NP21 (or *Ericsonia subdisticha* biozone) in the Northern Hemisphere as a function of latitude, in relation to the biozonation of planktonic foraminifera. The disappearance of the index species defining the upper and lower limits of the NP21 biozone appears earlier at high latitudes with the consequence that stratigraphic correlations established on the basis of this zone are invalid between tropical and northern domains. After Cavelier (1979, fig. 1)

appearance or disappearance of a species in one place to coincide with real emergence or extinction. More often, however, it is simply the result of migration. Moreover, biozones are bounded in space by the geographical extension of the index taxa none of which are of worldwide occurrence (Figs 15.3 and 4). Biostratigraphy is, in the end, no more than an ecostratigraphy. In practice, therefore, it has been necessary to draw up two zonations for each group of microfossils, one for tropical provinces and the other for boreal regions. Nevertheless, it remains true that different zonations with all their reciprocal relations have been found in basins that are far removed from each other. This is the important fact even though there may be minor anomalies here and there.

Homophyletic Biostratigraphy

An attempt has been made to eliminate the element of ecological bias by renouncing the use of zones based on biotic events and chosen arbitrarily from a succession of unrelated species. They are replaced by **lineage zones** or

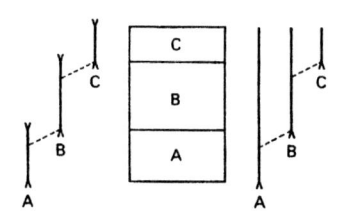

Fig. 15.5 Theoretical examples of phylozones: on the left, the zones are based on the lifespan of successive species of the same phyletic lineage; on the right, they are bounded by the successive appearances of different species of the same lineage

phylozones, which are based on the real appearance of species within a given phyletic lineage (Fig. 15.5). The lineages that are well known (like that which leads from *Globigerinoides* to *Orbulina*, cf. Fig. 13.18) are, however, rare. As their evolution has been followed in detail for only a short period, the method has, as yet, no real practical value. Nevertheless, it will be a path for future exploration that will be worth following from biological and geological viewpoints.

In the transition stage, a single characteristic intrinsic to one particular group is taken into consideration and its degree of evolution measured. By constructing and then standardizing a curve for this feature, it is possible to date isolated populations. The assumption is that the characteristic evolves irreversibly and at the same rate in every place. This seems to be the case for the embryo of the orbitolines of the Cretaceous (Fig. 13.5) and for that of the lepidocyclines of the Oligo-Miocene (Fig. 15.6).

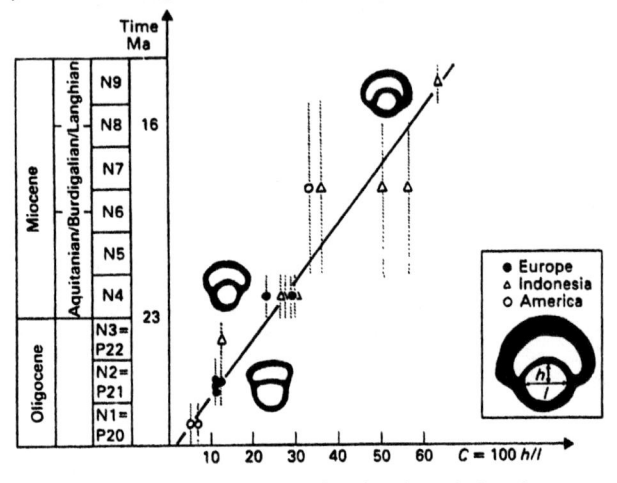

Fig. 15.6 An example of monophyletic dating based on the evolution of parameter C (= covering of the protoconch by the deuteroconch) in lepidocyclines from three different provinces in terms of ages obtained from the biozonation of planktonic foraminifera. After data in van Vlerk (1968)

3. BIOSTRATIGRAPHY, CHRONOSTRATIGRAPHY, MAGNETIC REVERSALS AND RADIOMETRIC DATING

The biozonations were originally set up more or less independently of the d'Orbigny chronostratigraphic scale, which, with numerous improvements, has served for a century as a universal framework for chronology. The fundamental unit in this framework is the stage, which takes as its standard form a **stratotype** or **series** of strata chosen from a particular region. Unlike the geographically bounded biozones, a stage includes, by definition, all the rocks formed during a given time interval. In order to establish correlations between the two scales, it is necessary to look for elements of comparability. One way of doing this is to see which index species for biozonations occur in the palaeontological content of the stratotype for each stage.

A similar method seeks to establish parallels between the zones and magnetic reversals, the latter phenomena having the advantage of being synchronous for the whole world.

Finally, figures have been given for the boundaries and duration of the subdivisions of the stratigraphic scale. However, it is far from simple to establish correlations between stratotypes and zonations on the basis of radiometric dating, which derives from the evidence of magmatic rocks and only very occasionally from sedimentary material (glauconite). Although controversy surrounds these datings, they allow estimates to be given for the duration of zones. The degree of precision varies with the rate of evolution of the group that is being considered. A zone may correspond to one or several stages; most commonly, five or six zones are distinguished for every stage. Such is the case for conodonts and chitinozoans in the Palaeozoic, and for planktonic foraminifera and coccoliths in the Cretaceous and the Cenozoic. The duraction of zones is in the order of a million years, like those based on trilobites, graptolites and ammonites.

It must be emphasized that biostratigraphic data, though relative, are more accurate than the findings of radiometry. Although the latter are quantitative, the degree of error that is involved (around 5%) is at least equal to the duration of a biozone.

The work of correlating biozones, stages, magnetic reversals and radiometric dating has been presented here in its logical order. It is to be regretted that this has not always been done elsewhere. Syntheses have been attempted prematurely in the past in the absence of sufficient analytical data. Those that are now being put forward are still working hypotheses (Fig. 15.7).

CONCLUSION

Microbiostratigraphy is a suitable method for dating sediments, especially when only small rock samples from cores, cuttings and undersea dredging are available. Nowadays dating is almost always on the basis of microfossils but in the past ages were attributed using macrofossils that were few in number, badly preserved or lacking in significance, or which derived merely from lithological and geometrical deductions. As a result these earlier dates have had to be revised.

It would be labouring the point to enumerate all the formations and even basins, formerly thought to be well known, whose ages have, in recent decades, been confirmed, corrected or simply established on the basis of microfossils. A single example will suffice: in the 300 m of white chalk of the Paris Basin, there are six to seven levels, characterized locally by echinoids and belemnites, which are difficult to pick out in the field because macrofossils are very uncommon. On older geological maps on the 1:80 000 scale, only two levels were

Chronostratigraphy			Planktonic microfossil biozonations		
Time (Ma) / Strato types	Stages		Foraminifera	Coccoliths	Dinocysts
33 37 —	late Eocene	Priabonian	P17 *Globigerina gortanii/ Globorotalia centralis*	NP21 *Ericsonia subdisticha*	W14 *Wetzeliella gochtii*
35 — 40			P16 *Cribrohantkenina inflata* (*Turborotalia cerroazulensis*)	NP20 *Sphenolithus pseudoradians*	W13
				NP19 *Isthmolithus recurvus*	*Kisselevia clathrata angulosa*
			P15 *Globigerinatheka semiinvoluta*	NP18 *Chiasmolithus oamaruensis*	
	Bartonian	Bartonian	P14 *Truncorotaloides rohri*	NP17 *Discoaster saipanensis*	W12 *Rhombodinium perforatum*
40 — 45			P13 *Orbulinoides beckmanni*	NP16 *Discoaster tani nodifier*	W11 *Rhombodinium porosum*
			P12 *Morozovella lehneri*		W10 *Rhombodinium draco*
	Lutetian	Lutetian	P11 *Globigerinatheka subconglobata*	NP15 *Chiphragmolithus alatus*	W9 *Wetzeliella articulata*
45 — 50			P10 *Hantkenina aragonensis*	NP14 *Discoaster sublodoensis*	W8 *Kisselevia fasciata*
					W7 *Kisselevia coleothrypta rotundata*
	Ypresian	early Eocene / Ypresian	P9 *Acarinina pentacamerata*	NP13 *Discoaster lodoensis*	W6
			P8 *Morozovella aragonensis*	NP12 *Marthasterites tribrachiatus*	*Kisselevia coleothrypta*
50 — 55			P7 *Morozovella formosa*	NP11 *Discoaster binodosus*	W5 *Dracodinium varielongitudis* W4 *D. similis* W3 *W. meckfeldensis* W2 *W. astra*
			P6 *Morozovella subbotinae*	NP10 *Marthasterites contortus*	
			P5 *Morozovella velascoensis*	NP9 *Discoaster multiradiatus*	W1 *Apectodinium homomorphum* (= *A. hyperacanthum*)
55 — 60	Thanetian	Thanetian	P4 *Planorotalites pseudomenardii*	NP8 *Heliolithus riedeli*	*Deflandrea speciosa*
				NP7 *Discoaster gemmeus*	
			P3b *Planorotalites pusilla*	NP6 *Heliolithus kleinpelli*	
				NP5 *Fasciculithus tympaniformis*	
60 — 60		Palaeocene	P3a *Morozovella angulata*	NP4 *Ellipsolithus macellus*	
	Danian	Danian	P2 *Morozovella uncinata*	NP3 *Chiasmolithus danicus*	*Paleoperidinium basilium*
			P1d *Subbotina trinidadensis*		
			P1c/b *Subbotina pseudobulloides*	NP2 *Cruciplacolithus tenuis*	
65 65			P1a *'Globigerina' eugubina*	NP1 *Markalius inversus*	

distinguished. On recent 1:50 000 maps, C. Monciardini has subdivided the same chalk into ten biozones on the basis of small benthic foraminifera. This improvement has led to a better reconstruction of the geological history of the basin as the detailed cartography of the ten zones shows up sedimentary gaps and concentrations as well as revealing an ante-Eocene tectonic deformation.

At present, an effort is being made to extend the microbiostratigraphic method by applying it to metamorphic rocks containing radiolarians, conodonts and dinocysts. Although there have been difficulties and disappointments, unexpected and highly significant results have been obtained. To take only one example: the recent discovery by Ch. Ioakim of Oligocene dinocysts in the phyllades of the Peloponnese has obliged specialists to take a new look at tectonic models for the Greek peninsula.

Although restricted in space and variable in duration, the biozonations, particularly those based on planktonic microfossils, form the best available method at present for dating Phanerozoic sedimentary strata in terms of ease, rapidity, cost, applicability and accuracy. The method is continuously being improved. Recent research is showing that certain groups, such as radiolarians and diatoms, which were considered a few years ago to be of no stratigraphic value are in fact excellent chronometers.

With all its diverse possibilities, the microbiostratigraphic method is now a well-honed tool that can be used with great precision. It is organized on an industrial scale and is in constant demand whenever sedimentary rock is being examined for useful materials (e.g. coal, oil, some ores, underground water) or for the construction of civil engineering works (e.g. tunnels, ports, dams, power stations).

BIBLIOGRAPHY

The different methods used in stratigraphy are treated in the reports of the *Colloque sur les Methodes et Tendances de la Stratigraphie (Orsay, 1970)*, in *Mémoires du Bureau de Recherches géologiques et minières* no. 77 (1972; 2 vols) and in Ch. Pomerol, Cl. Babin et al., *Stratigraphie et Paléogeographie: Principes et Méthodes (Doin, Paris, 1980)*.

It is impossible to choose among the innumerable works dealing with the contribution of microfossils to the dating of sediments.

Microfacies are illustrated in several publications, notably those of J. Cuvillier, *Stratigraphic Correlations by Microfacies in Western Aquitaine* (Brill, Leiden, 1956); D. Tedeschi, *Microfacies italiane* (Agip. Min., San Donato Minanese, 1959); and V. Sacal & J. Cuvillier, 'Microfacies du Paléozoique saharien', *C.F.P., Notes & Mémoires* 6, (1963). Also, consult R.M. Stainforth et al., Cenozoic planktonic foraminiferal

---

Fig. 15.7 An example of correlation between some interrelated biozonations (biostratigraphy), stage stratotypes (chronostratigraphy) and radiometric ages (Ma) suggested for the Palaeocene and Eocene. Two different series of radiometric datings are in favour among specialists. The first of these (1) is from Odin, Curry & Hunziker (1978). The other (2) derives from Hardenbol & Berggren (1978). Only three biozonations have been taken into account. It would have been useful to complete the table by the addition of other planktonic microfossil biozonations (e.g. radiolarians, silicoflagellates, diatoms), or even benthic microfossil biozonations (e.g. *Alveolina*, *Nummulites* and *Assilina*). After Bignot & Cavelier (1981).

zonation and characteristics of index forms, *University of Kansas Palaeontology Contribution*, (No. 62 (1975), and W.W. Hay, *Calcareous Nanofossils* (op cit.)).

Planktonic microfossils and the biozones that they define are abundantly described in the *Proceedings of the International Conference on Planktonic Microfossils. (Geneva, 1967*, publ. 1969, Brill, Leiden, 2 vols; and Rome, 1970, publ. 1971, Tecnoscienza, Rome, 2 vols).

Some idea of the importance of the contribution of micropalaeontology to ocean drilling can be gleaned from the *Initial Reports of the Deep Sea Drilling Project* (since 1968 more than 100 vols each of around 1000 pp.); J.E. Warme, R.G. Douglas & E.L. Winterer (eds), 'The Deep Sea Drilling Project a decade of progress', *Special Publications of the Society of Economic Paleontology and Mineralogy* 32 (1981).

For general references on micropalaeontological biostratigraphy see the following: F.P. Agterberg, and F.M. Gradstien, Workshop on "Quantitative Stratigraphical Correlation Techniques", Ottawa 1980. (*Int. Assoc. Math. Geol. J.* 13, 81–91); W.H. Berger and E. Vincent, "Chemostratigraphy and biostratigraphic correlation; exercises in systemic stratigraphy". (*Oceanol. Acta*, 1981, 115-127); R.G. Blank and G.H. Ellis, "The probable range concept as applied to the biostratigraphy of marine microfossils", (J. Geol. 90, 1982, 415–433); A.E. Cockbain, "An attempt to measure the relative biostratigraphic usefulness of fossils", (*J. Paleontol. 40* (1966), 206-7); C.W. Drooger, "The boundaries and limits of stratigraphy", (*Koninkl., Nederl., Akad., v. Wetens – Amsterdam*, 1974, Proc. (B) 77 (3) 159–176); H.D. Hedberg (Ed), *International Stratigraphic Guide* (John Wiley, 1976); R.J. Price and P.R. Jorden, "A FORTRAN IV program for foraminiferal stratigraphic correlation and palaeoenvironmental interpretation", (*Comput. Geosci.* 3 (1977) 601-615).

Chapter 16

Microfossils as Palaeoenvironmental and Palaeogeographic Indicators

No matter what the size of the fossils, the same procedures must be followed whenever palaeoenvironments and palaeogeographies are reconstructed from palaeontological data. The treatment that follows, therefore, will be no more than a review of general principles. On the other hand, considerable space is devoted to the presentation of results obtained recently through the study of microfossils.

1. FROM PALAEOECOLOGY TO THE RECONSTRUCTION OF PALAEOENVIRONMENTS

General Principles

It has already been seen that:

- every modern biotope is characterized qualitatively and quantitatively by particular biocoenose; and
- it is possible to determine the mode and conditions of life of fossils.

That implies that it is possible, from palaeoecological consideration, to reconstitute some of the characteristics of paleoenvironment. However, there is a serious difficulty that has already been referred to: it is not biotopes or fossil biocoenoses that are being analysed but sediments and taphocoenoses. Only through these can access to different palaeoenvironments be obtained. Sometimes, the palaeoecological data are sufficient to distinguish a biocoenose from a taphocoenose. The foraminifera that are found in oceanic sediments are 99% planktonic. As only the remaining 1% of benthic forms live on the ocean floor, it follows that all the others have come from the superjacent waters.

Most often, however, the taphocoenose is considered as a whole and the evidence that it presents is neither clear nor simple. Local distributions and minor modifications in time cannot be detected. It is only general trends valid for the region that are accessible.

Benthic environments are preserved by an abiotic phase, the sediment, and a biotic phase, the taphocoenosis. For certain other aquatic and subaerial environments, however, the only evidence is the biotic phase. Knowledge of oceanic waters can be obtained only through nectonic microfossils (e.g. ammonites, graptolites and fish) and planktonic microfossils (e.g. pteropods, foraminifera, coccoliths and radiolarians). Subaerial terrestrial environments are manifested in marine sediments by microfossils carried in water (spores, charophyte gyrogonites) or in the air (pollens).

Fossils are, therefore, the bearers of valuable information for the reconstruction of palaeoenvironments. Such data must, of course, be checked against information from other sources that have not been dealt with here: for example, the cartography, lithology, sedimentology and geochemistry of the deposit. It should also be noted that the state of preservation of the fossils and their distribution in the sediment provide information on the chemistry of the environment of fossilization and on the dynamics of the surrounding waters.

Palaeoecological Procedure

It is usually possible to draw conclusions from a simple glance at the qualitative composition of a taphocoenosis if this contains adequate material; such is generally the case with microfossils. This empirical procedure is based on the criteria of the presence or absence of genera or higher groups of known palaeoecology whose existence in the period under consideration is established.

The presence or absence of representatives of stenohaline groups (radiolarians, planktonic foraminifera on the one hand, and charophytes on the other) immediately gives the salinity of the waters or – at the very least – differentiates marine from freshwater environments. Low salinity ($<5‰$) is revealed by an abundance of ostracods with a tuberculate carapace (e.g. Purbeckian and Wealdian facies containing *Cypridea*) or by large mono- or oligospecific populations (e.g. Bartonian facies of the Paris Basin containing *Rosalina bractifera* and *Cypris tenuistriata*).

Shallow water (a few score metres at most) is indicated by the presence of red or green calcareous algae which are attached to the sea-floor and are confined to the euphotic zone. It is, however, also necessary to ascertain that movement before burial was not great. Apart from littoral environments, it is always difficult to measure palaeodepths with accuracy. For this reason, it is usually preferable to estimate the distance from the environment to the coast and the extent of communication with the open sea. These two factors, which are generally but not always in contrast to each other, are linked in the concept of **oceanicity** (and its opposite, **confinement**): an oceanic environment may be close to the coast where the shelf is narrow and, conversely, a gulf, which reaches far inland, may receive plankton through the movement of currents. The more numerous the microplankton, the greater the degree of oceanicity.

Palaeotemperatures can be deduced directly from the palaeoecological data

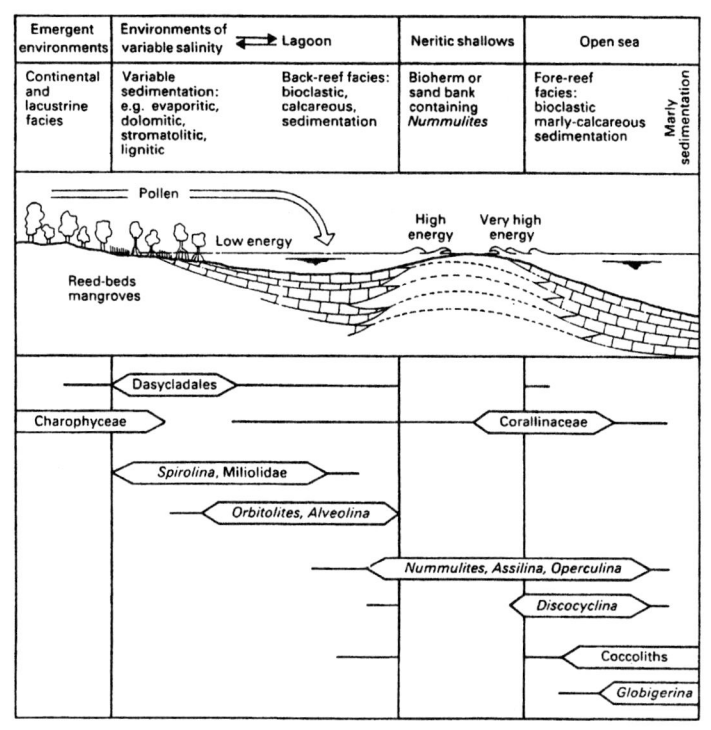

PALAEOENVIRONMENTAL AND PALAEOGEOGRAPHIC INDECATORS

Emergent environments	Environments of variable salinity ⇌ Lagoon		Neritic shallows	Open sea	
Continental and lacustrine facies	Variable sedimentation: e.g. evaporitic, dolomitic, stromatolitic, lignitic	Back-reef facies: bioclastic, calcareous, sedimentation	Bioherm or sand bank containing *Nummulites*	Fore-reef facies: bioclastic marly-calcareous sedimentation	Marly sedimentation

Fig. 16.1 Ecozonation of Mesogean platforms in the Eocene (around 40 to 50 million years ago). After Arni (1965, fig. 2)

as certain species are typical: the presence of large foraminifera with many symbiotic zooanthellae indicates temperatures higher than +18 °C as well as shallow water. Microplankton are also good indicators of temperature (Fig. 13.7). For any given period the population stages define a (palaeo)ecological zonation or ecozonation. P. Arni's model, which is summarized in Fig. 16.1, concerns only foraminifera and a few other microfossils of the Eocene, but with slight changes it is applicable to the whole of the Cenozoic. It makes it possible, with only a brief examination, to give the approximate position of any deposit, to fix its place in the succession of palaeoenvironments and to evaluate from this the depth, salinity and dynamics of the water.

Palynological analysis similarly makes it possible to reconstruct the appearance of plants with greater accuracy than is the case using the 'macroremains' of flora. On this basis, C. Gruas-Cavagnetto (1968) has shown that during the brief Sparnacian lacustrine episode (early Eocene), there coexisted in the Paris Basin:

- A vegetation floating on the surface of the water, consisting of *Botryococcus* and primitive Chlorophyceae, Sparganiaceae and Hydrofilicales (e.g. *Azolla* and *Salvinia*).
- A hygrophil association, bordering marshes and lowlands subject to flooding, with *Lycopodium*, herbaceous ferns (e.g. Schizaeaceae, Polypodiaceae, *Osmunda*), palms (e.g. *Sabal*, *Calamus* and *Nipa*) and Taxodiaceae.
- An evergreen forest with *Platycarya* and other Juglandaceae, together with Ebenaceae, Hamamelidaceae and Ulmaceae, covering an undulating hinterland in much the same way as the forests of southern China today.

Wherever fossils are abundant, statistical analyses are possible and, in practice, microfossils are frequently used, particularly palynomorphs, foraminifera and ostracods. Fruitful comparisons can be drawn between fossil deposits and the thanatocoenoses found in modern environments. Although the results may concern only one particular deposit, they gain from being set in the context of space and/or time. The evolution of one or several ecological factors is then summarized in the form of a map or log. The latter is often used as it indicates

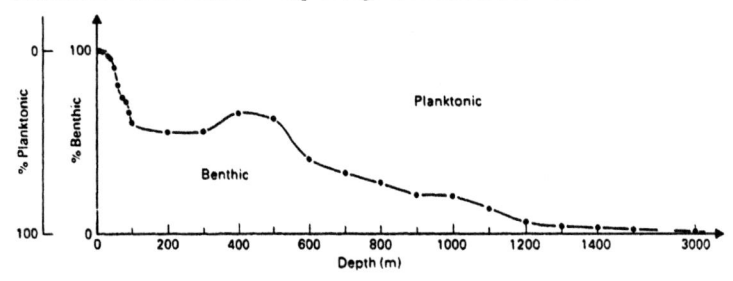

Fig. 16.2 Evolution of the ratio of planktonic to benthic foraminifera as a function of depth in the Gulf of Mexico. After Grimsdale & van Morkhoven (1955, fig. 1)

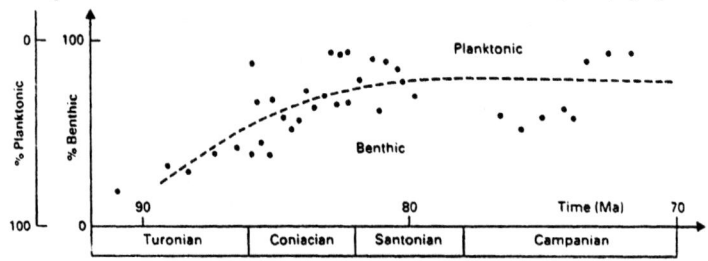

Fig. 16.3 Evolution of the ratio of planktonic foraminifera to benthic foraminifera in the late Cretaceous chalks of the Anglo-Parisian Basin (cliffs of the Isle of Wight and Haute-Normandie). From the numerical data in the preceding curve, it can be estimated that, in the Turonian (around 90 million years ago), the depth of water covering the basin was between 600 and 800 m. This value subsequently diminished to stabilize at around 100 m from the Santonian (80 million years ago). After Barr (1962, text-fig. 3) and Chemirani (1968, table 12)

clearly the significance of the results obtained from drillings. The degree of oceanity and the bathymetry of ancient basins can be assessed with the aid of values for the dinocyst/pollen ratio or the ratio for planktonic to benthic foraminifera in the sediments of present-day seas (Figs 16.2 and 3).

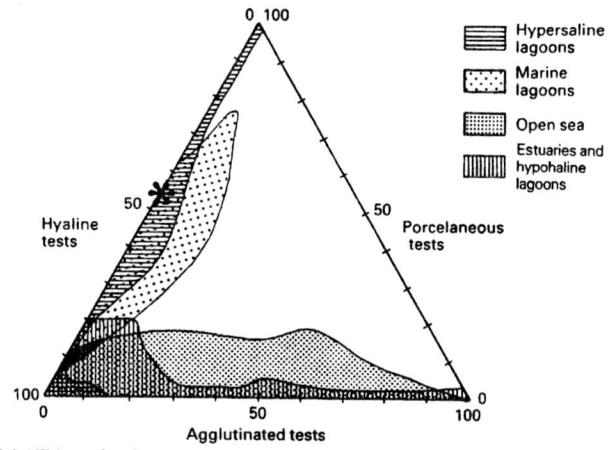

Fig. 16.4 Triangular diagram in which the emplacement of a deposit is fixed by the relative proportions in the taphocoenose of benthic foraminifera of different test types (agglutinated, porcelaneous, hyaline). The areas delimiting the different facies have been established from biocoenoses found in nature today. The famous Lutetian deposit at Grignon to the south of Paris (indicated on the graph by a star) corresponds to an ancient lagoon which was either marine or slightly hypersaline. After Murray (1973, fig. 102)

The diversity expressed in J.W. Murray's triangular diagram (Fig. 16.4) makes it possible to estimate the values for palaeosalinity and to compare a fossil environment with its present equivalent in terms of salinity and geography.

Palaeotemperatures can be assessed by determining the respective percentages for two species of silicoflagellates, *Dictyocha octonaria* for warm waters and *Distephanus speculum* for cold, and, even more importantly, the ratio of dextral to sinistral forms in the same species of planktonic foraminifera. These figures are then compared with the corresponding values in modern oceans. Another procedure that produces clear indications is based on shifts in latitude for microfossil populations in relation to climatic fluctuations. The Quaternary is particularly suitable for this method as the plant and animal associations are the same as those of today. Local evolutions in climate can be reconstructed from variations in the abundance of ecomorphs (Fig. 16.5), species and populations (Fig. 16.6) as compared with their present distribution. At any given point, these variations mark biogeographical movements in the course of time. When, for example, extensions of known periods for the biocoenoses of some planktonic foraminifera are transferred to the map, it is revealed how cold Arctic waters have moved towards the south (Fig. 16.7).

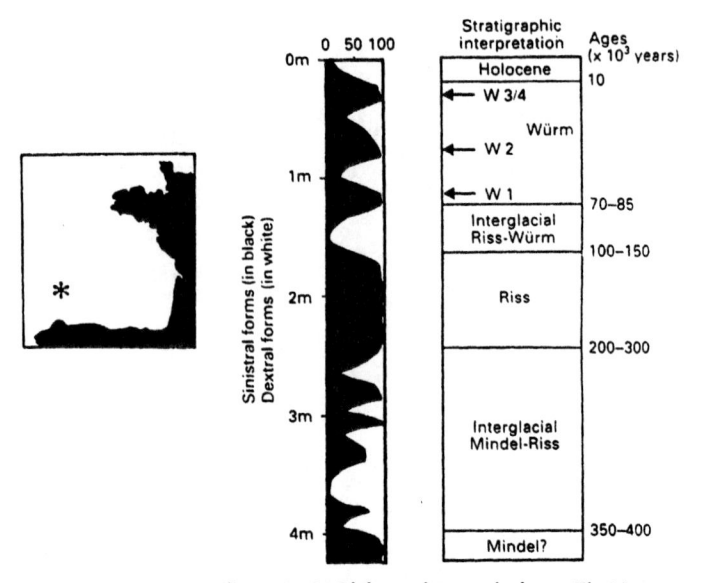

Fig. 16.5 Variations in coiling ratios in *Globoquadrina pachyderma* (Ehr.) in a core sample of Quaternary sediments from the Gulf of Gascony off Corunna. The predominance of sinistral forms corresponds in each case to a cold period (cf. Fig. 13.16). After Caralp & Pujol (1974, fig. 35)

The same method has been used to reconstruct the forest history of the Holocene (Fig. 16.8) and to trace the essential elements in human life through deforestation, pasturage, and the introduction of cultivated plants.

Shifts in latitude for more recent periods are more difficult to establish for the following reasons:

- Recent species do not appear before the Neogene. For earlier periods, it is necessary to determine first of all the natural associations and then to establish their climatic preferences.
- Latitudinal zonation – as will be shown further on – becomes less clear the further back it is traced through geological time.

Significant results have, nevertheless, been obtained (Fig. 16.9).

The preceding examples, both qualitative and quantitative, relate to environments with an age of less that 100 million years. The further back data from the present are extrapolated into geological time, the less they can be relied on. The ratio of benthic to planktonic foraminifera, for example, can have little meaning before the late Cretaceous as the planktonic species did not appear in the Jurassic and it was not until the Cretaceous that their development could be compared with that of today. The older a taphocoenosis, the more problematic its interpretation and the more dubious its value for reconstructing palaeoenvironments. Beyond the Jurassic (>150 to 200 million

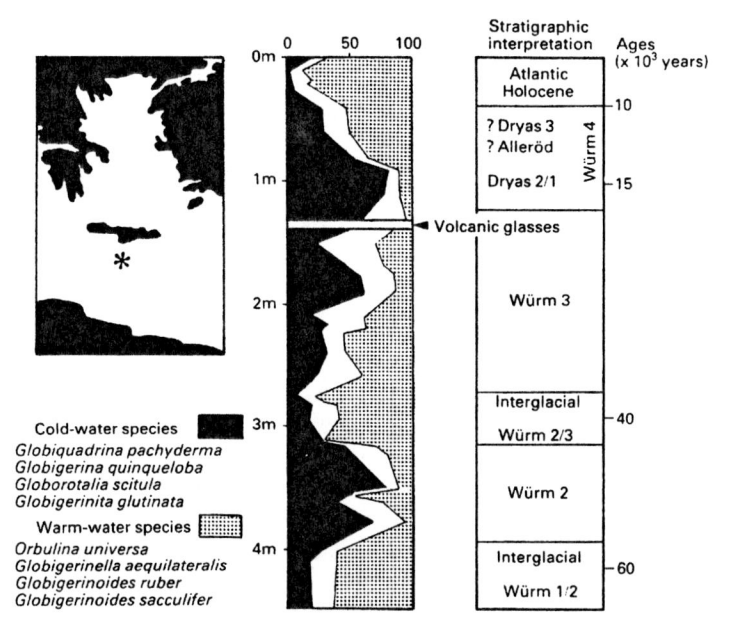

PALAEOENVIRONMENTAL AND PALAEOGEOGRAPHIC INDECATORS 193

Cold-water species
Globiquadrina pachyderma
Globigerina quinqueloba
Globorotalia scitula
Globigerinita glutinata

Warm-water species
Orbulina universa
Globigerinella aequilateralis
Globigerinoides ruber
Globigerinoides sacculifer

Fig. 16.6 Variations in the relative importance of assemblages of planktonic foraminifera in boreholes in Quaternary sediments to the south of Crete. After Blanc-Vernet (1972, fig. 3)

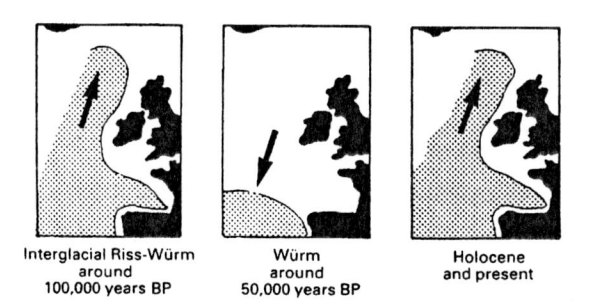

Interglacial Riss-Würm around 100,000 years BP

Würm around 50,000 years BP

Holocene and present

Fig. 16.7 Extension in the north-west Atlantic Ocean of two warm-temperate species of planktonic foraminifera: *Globorotalia hirsuta* (D'Orb.) and *G. truncatulinoides* (D'Orb.) during the last three climatic episodes. After Caralp & Pujol (1974, fig. 12)

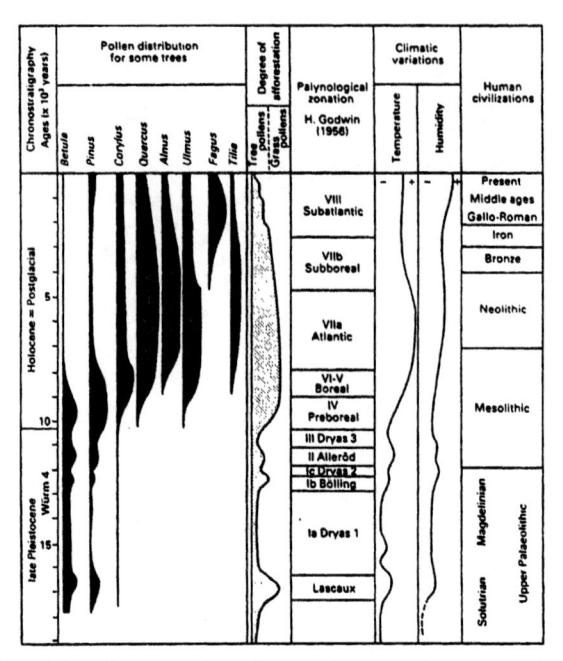

Fig. 16.8 Evolution of vegetation in northwestern France during the past 20 000 years. The extremely cold climate of the Würm brought a treeless tundra or taiga with stunted birches. Increasing warmth and precipitation in the postglacial period allowed pine forests to become established followed by oaks and finally beeches. From the degree of afforestation, it can be seen that land clearance began in the Neolithic and accelerated in the Roman period. Largely after data in Elhai (1963)

years), the palaeoecological method gradually loses credibility and it becomes necessary to rely on sedimentary data for such reconstructions.

There is, in fact, no single method of reconstruction but rather a general principle based on the postulate that 'the present is the key to the past' and this is used in a multitude of more or less empirical procedures. Despite, or perhaps because of, that fact, this line of research is, for the present, the most original and the most fruitful that is open to micropalaeontologists.

Isotopic Determination of Palaeotemperatures

This method, which stems from the work of H.C. Urey (1947), is not, properly speaking, micropalaeontological. However, it is worth a brief mention here as the material that is used is, in practice, almost always formed from microfossils.

Natural oxygen is a mixture formed mainly from two stable isotopes, with masses of 16 and 18, in the proportion:

$$\frac{^{18}O}{^{16}O} = \frac{1}{500} = 0.002$$

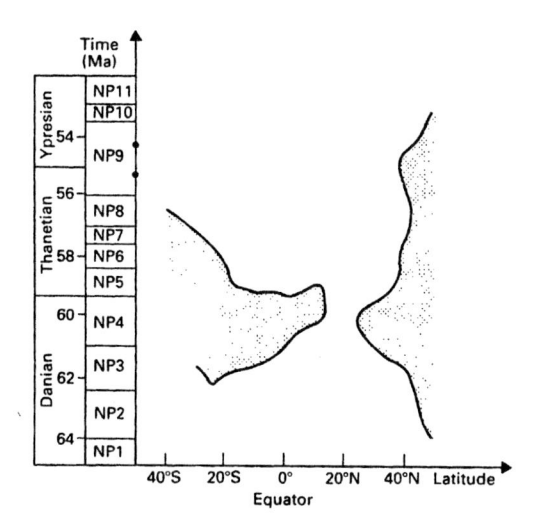

Fig. 16.9 Distribution in time and space (shaded area) of the coccolith assemblage, *Prinsius martinii* during the Palaeocene. This bipolar assemblage appeared around 64 million years (Ma) ago in high latitudes and then migrated towards the Equator around 60 million years ago. It returned to the high latitudes subsequently and finally disappeared around 53 million years ago. It is thought that the assemblage was restricted to temperate waters and that its temporary migration towards the Equator was linked with a cooling of the surface waters of the ocean. After Haq, Okada & Lohmann (1979, fig. 6)

The oxygen of calcium carbonate has an isotopic composition (d_c) slightly different from that (d_w) of the surrounding water. Isotopic analysis of the oxygen in the shells of molluscs raised to controlled temperatures has shown in experiments that the temperature of the water in the biotope is given by the simplified formula:

T (°C) = 16.9 − 4.2 ($d_c - d_w$)

The d_c is measured in relation to the isotopic composition of a standard carbonate termed PDB (= Pee Dee belemnite)

d_c or $\delta = 1000 \; \dfrac{^{18}O/^{16}O \text{ sample}}{^{18}O/^{16}O \text{ standard PDB.}} - 1$

To calculate T from d_c, d_w must be constant with the passage of time. However, the evaporation of $H_2{}^{16}O$ takes place more rapidly than that of $H_2{}^{18}O$ as it is lighter. Atmospheric water vapour deriving from the evaporation of the oceans is thus richer in ^{16}O. The accumulation of this vapour on the continents in the form of ice is accompanied by a corresponding increase of the ^{18}O content of the oceans. It is impossible to give an accurate figure for the increase in d_w as this depends on the quantity of ice trapped by the poles. For this reason, T cannot be calculated for periods before the formation of the present ice-caps (i.e. 2.5 million years for Greenland and between 11 and 14 million years for the Antarctic). Before these dates, the

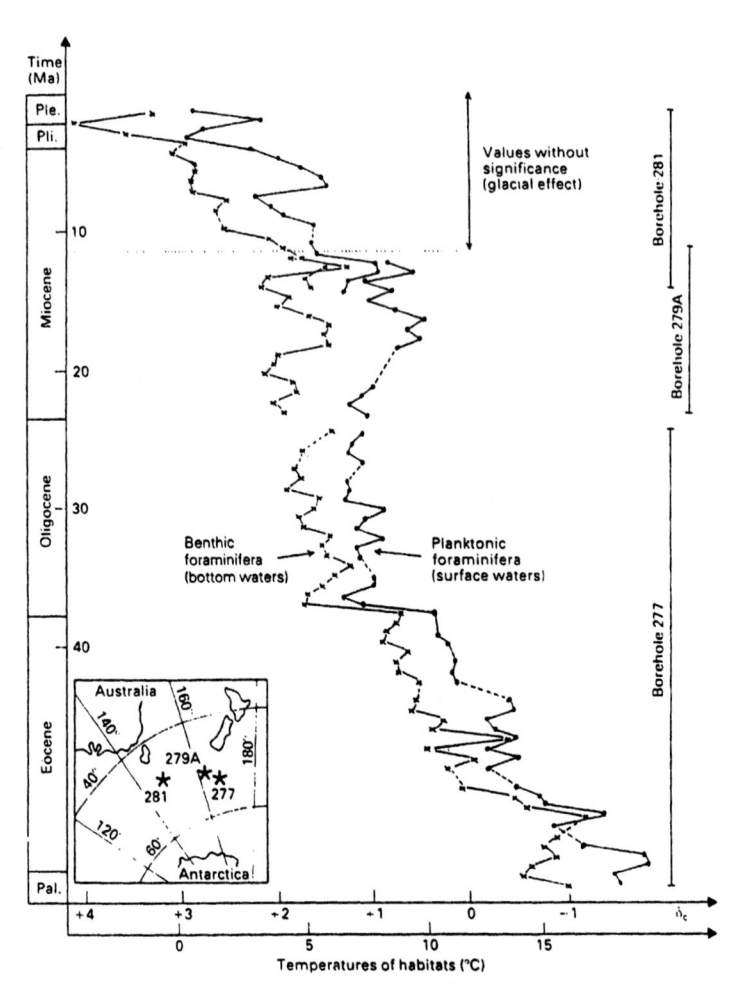

Fig. 16.10 Evolution of the isotopic oxygen composition of the tests of benthic and planktonic foraminifera. The tests derive from a continuous sedimentary series lying between Australia and Antarctica and are taken from three boreholes the positions of which are indicated in the box at bottom left. The temperature in the surface habitat of planktonic foraminifera is some 3 to 4 °C higher than that of the deep habitat of benthic foraminifera. Temperatures fall throughout the Eocene, undergo an abrupt drop at the Oligocene–Eocene boundary and finally stabilize. The values for the past 11 million years ago are without significance as d_w varies with the quantity of ice accumulated at the poles. After Shackleton & Kennett (1975, fig. 3)

average isotopic composition of seawater must have been 1.2% lower than that of the PDB standard.

Isotopic analysis can be carried out on any carbonate. Preliminary examinations for diagenetic modification, impurities, etc. are extremely important but this is not the place to enumerate them. The biogenic carbonates that are most favoured for use are the tests of foraminifera. In practice, 5 mg of carbonate, representing several dozen tests, provides sufficient material. Variations in the d_c of foraminiferan tests (Fig. 16.10) successfully illustrate the gradual but perceptible cooling of the oceanic environment. Periods of accelerated cooling are also shown, the main one being situated at around 38 million years on the boundary between the Eocene and the Oligocene.

Palaeotemperatures have also been calculated on the basis of the isotopic oxygen composition of the silica in radiolarians and diatoms.

Finally, the isotopic composition of the carbon ($^{13}C/^{12}C$) in the carbonates of foraminiferan tests has also been used though its interpretation remains difficult. It would seem to show the interaction of photosynthesis in surface waters and the oxidation of organic matter in deep waters. For this reason, the method should permit the location of fossil zones in which there is high organic productivity.

2. MICROFOSSILS – EVIDENCE OF SEA-FLOOR SPREADING

Ecology and Biogeography

For every species, there should be a suitable habitat within the boundaries of some potential distribution. In fact, however, the areas of actual distribution are invariably smaller, some being very restricted and others discontinuous. When the areas of real distribution for numerous species are placed on the same map, it becomes possible to define biogeographic provinces or bioprovinces (Figs 13.6 and 13.7). These lie roughly parallel to the Equator and are separated from each other by topographical, hydrological and thermal barriers, which are obstacles in the way of the dispersion of species.

Bioprovinces and Movements of Continents

Already in 1912, A. Wegener had pointed out that southern Africa, Madagascar, Australia, Antarctica and South America were covered with a uniform vegetation from the end of the Palaeozoic to the Jurassic. To explain how this palaeoflora had come to be divided up into regions now far removed from each other, he postulated the existence of a supercontinent, Gondwana, which broke up during the Mesozoic into fragments that drifted apart.

These observations were confirmed by micropalaeontologists. In the regions of Bahia and Sergipe of eastern Brazil and along the coast of Gabon, there are outcrops of lacustrine sediments dating from the early Cretaceous (Wealdian). Despite the distance that now separates them, both yield microflora and assemblages of ostracods with many species in common (Fig. 16.11). It is

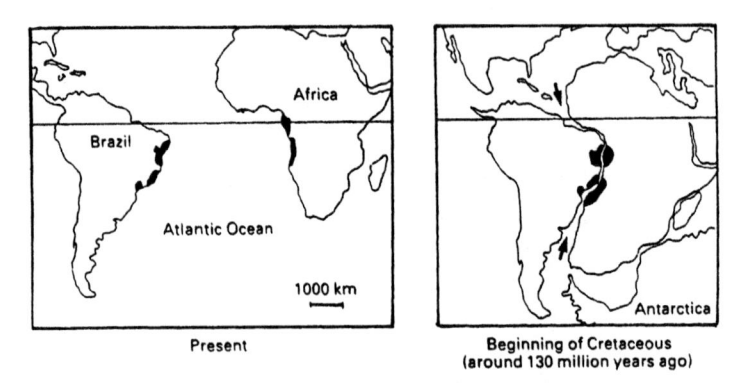

Fig. 16.11 Present position (on the left) and position at 130 million years ago (on the right) of the Wealdian sediments of Brazil and Central Africa. They were laid down initially in lakes along the continental rift from which the Atlantic Ocean was to spring

inconceivable that the remains of such terrestrial and lacustrine organisms could have been disseminated in a way so that they became distributed on either side of a 4000-km wide ocean. On the other hand, the observations conform with the hypothesis of continental drift. Furthermore, the marine sediments of the late Cretaceous overlying the Wealdian in both Brazil and Gabon, show similarities of microflora and microfauna which subsequently are less clear. Deriving from a common stock, the Brazilian and African populations had a parallel evolution on the continents and in the coastal seas but gradually diverged from each other with the opening up of the Atlantic Ocean.

Microfossils, Magnetic Anomalies and Sea-floor Spreading

The most striking development in recent years has probably been the part played by micropalaeontology in confirming the hypothesis of sea-floor spreading, which, historically, is one of the fundamental concepts of plate tectonics. The facts may be summarized as follows: the ocean floor contains some 200 linear magnetic anomalies parallel to the Mid-Atlantic Ridge and these are believed to result from the fossilization of the Earth's magnetic field at the time of emplacement of the corresponding layer. By correlating the linear anomalies closest to the ridge with reversals known on land, J.R. Heirtzler and his collaborators deduced the rate of sea-floor spreading for the past 5 million years. A bold extrapolation of this supposed rate allowed them to date the 34 most recent anomalies within 80 million years at most.

A vast programme of ocean drilling (DSDP = Deep Sea Drilling Project) was set up in 1968 to attempt to verify the hypothesis. Initially, the project was entirely American (JOIDES = Joint Oceanographic Institutions Deep Earth Sampling) but, since 1975, it has been international (IPOD = International Phase of Ocean Drilling). The programme continues and is already paying dividends in the form of the 500 or more drillings carried out by the exploration ship *Glomar Challenger*.

Several drillings were carried out on linear anomalies that had been precisely identified. The core samples taken from beneath the sediments consisted of altered basalt that does not lend itself to radiometric dating. Its emplacement is supposed to be immediately anterior to the directly superjacent sediment. The sediment has been dated through planktonic microfossils (foraminifera and particularly coccoliths) and the values obtained differ by only 12% from those proposed by Heirtzler (Fig. 16.12). The oldest anomalies (M1 to M25) were dated from 120 to 150 million years exclusively on the basis of microfossils.

The hypothesis of sea-floor spreading has, thus, been verified by micropalaeontologists. They, for their part, have obtained a better understanding of palaeobiogeography and of the possibilities that exist for analysing the palaeontological content of the exceptional sedimentary series being sampled.

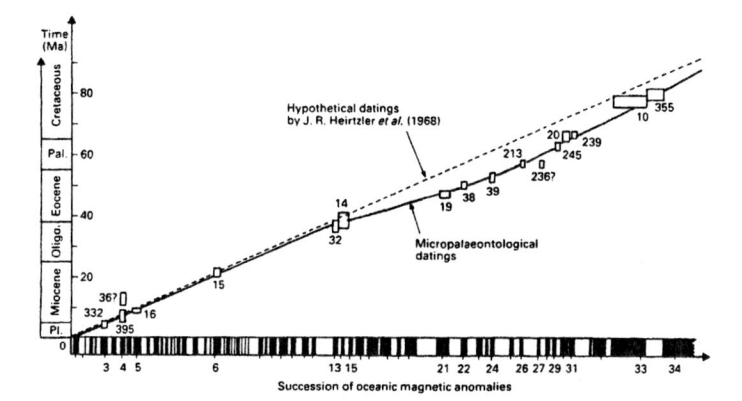

Fig. 16.12 Dating of linear magnetic anomalies on the ocean floor by:
• Extrapolation from the first four reversals in the scale of magnetic polarity (hypothesis of Heirtzler, Dickson, Herron, Pitman & Le Pichon, 1968)
• Biozonations of planktonic foraminifera and coccoliths and the corresponding radiometric datings (data taken from numerous publications, including Schlich, Muller & Sigal, 1979)

The numbers figuring on the curve correspond to sites drilled by the DSDP. At these sites, chosen for their known relationship to the linear anomalies, it was possible to obtain micropalaeontological dating of sediments lying directly on the oceanic crust.

The variation in size of the rectangles shows the uncertainty in dating (vertical direction) and the uncertainty connected with the relationship to linear anomalies (horizontal direction). The succession of magnetic anomalies is established on the basis of a constant rate of expansion for the sea-floor without reference to real distance from the mid-oceanic ridge. The distances, in fact, vary according to the ocean. It should be noted that, up to anomaly 13, there is a close correspondence between the datings proposed by Heirtzler and those established on the basis of microfossils. Beyond this point, the curves part company. The discrepancy reaches 7 to 8 million years ago between anomalies 25 and 26, which suggests variations in the rate of spreading

200   PART 2: GEOLOGICAL AND PALAEOBIOLOGICAL APPLICATIONS

Studies are still not completed but the results obtained have changed the picture in many fields, particularly taphonomic phenomena, speciation, modes of evolution and the use of microfossils in stratigraphy.

To sum up, the dating of the ocean floors, from which the history of the oceans and the successive displacements of the continents may be deduced, can be tackled by two methods:

- the distribution of linear magnetic anomalies and
- micropalaeontological datings.

The former, however, cannot be fully accepted without confirmation from the latter. This is not the least of the successes achieved by the study of microfossils.

3. FROM PALAEOBIOGEOGRAPHY TO GLOBAL PALAEOGEOGRAPHY

Microfossils and Palaeogeographic Changes of the Tethys and the Atlantic

The bioprovinces revealed by emergent sediments have been compared with the fragmentation and lateral displacement of continental masses during the past 200 million years. The distribution of marine microfossils, especially foraminifera, is reasonably consistent with the subduction of the Tethys and the opening of the Atlantic.

The microfauna of the Jurassic is distributed according to a geography quite different from that which exists today. Radiolarians, coccoliths, calpionellids, *Nannoconus* and planktonic foraminifera flourished in the surface waters of the Tethys, an ocean open towards the east and driven like a wedge into the enormous continental mass of Pangaea (Fig. 16.13). From Mexico to Iran, the continental shelf contained populations of large lituolid foraminifera. Parallel to this, on either side of the Tethys, the neritic seas, temporarily invaded by plankton, contained permanent populations of small benthic foraminifera. These were so widespread that they occur in North America, Europe and even

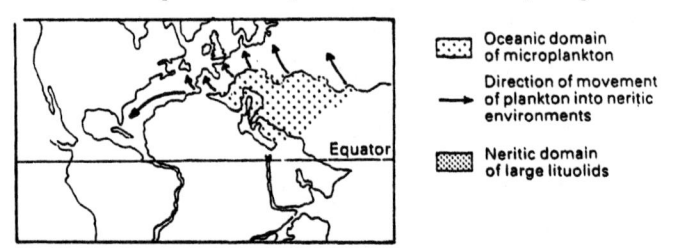

Fig. 16.13 Distribution of associations of foraminifera in the Tethys Ocean during the late Jurassic (around 150 million years ago)

Australia and New Zealand. Determined by the palaeogeography, the distribution of these microfossils shows no sign of division into provinces whether by latitude or longitude, a situation that persisted until the beginning of the Cretaceous when the same large benthic foraminifera occur from Mexico to the Middle East.

It is only from the late Cretaceous that the trend towards the formation of provinces appears. Some small benthic foraminifera (*Bolivinoides*) and the orbitoidids (*Orbitoides*) remain ubiquitous. The alveolinids (e.g. *Ovalveolina* and *Praealveolina*) and the lituolids (*Cyclolina*), however, though abundant around the Mediterranean and in the Middle East, are lacking in the Caribbean where the pseudorbitoidids are dominant.

This emergence of a longitudinal province is a sign of the opening of the Atlantic, and to it must be added latitudinal provinces attested by the distribution of coccoliths and planktonic foraminifera. Certain genera (*Praeglobotruncana*) are ubiquitous; others (e.g. *Pseudotextularia* and *Globotruncana*) prefer warm waters and still others (e.g. *Heterohelix*, *Hedbergella* and *Globigerinelloides*) cold waters (Fig. 16.14). The establishment of temperate and tropical oceanic bioprovinces is indicative of the end of climatic uniformity and the beginning of a latitudinal thermal gradient. The fact that the tropical province extends northwards to the coasts of Europe probably indicates the existence of a current (proto-Gulf Stream) flowing from the south-west.

This double evolution was continued during the Cenozoic. The division into latitudinal provinces was reinforced with the separation of Spitzbergen from Greenland and the consequent entry of cold Arctic waters into the Atlantic. The distribution of coccoliths and planktonic foraminifera tended gradually towards the arrangement that exists today. The extraordinary development

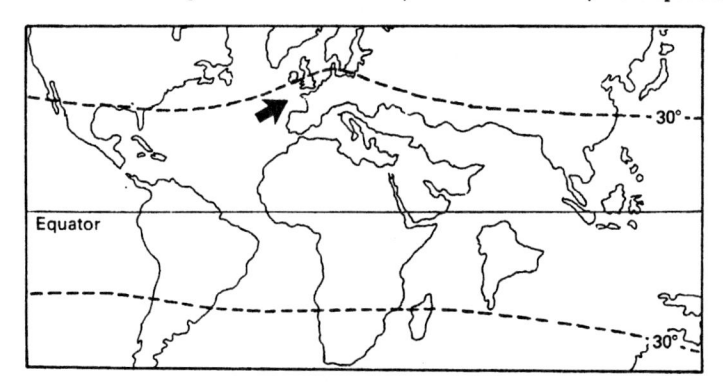

Fig. 16.14 Latitudinal provincialism of coccoliths and planktonic foraminifera at the end of the Cretaceous (around 63 million years ago). The warm water associations are concentrated roughly between latitudes 30°N and 30°S. The northward bulge (indicated by the arrow) showing the spread of tropical associations to the coasts of Europe points to the existence of a proto-Gulf Stream coming from the south-west (cf. Fig. 13.6). After Worsley & Martini (1970)

around the Mediterranean of alveolinids, orbitolitids and *Assilina* stands in contrast to their absence from the Caribbean where they were replaced by other foraminifera (*Yaberinella* and helicolepidinids). Although the Atlantic was progressively opening, it seems that it could still be crossed as the discocyclinids and some *Nummulites* of Mesogaean origin managed to reach the Caribbean. In the reverse direction, the lepidocyclinids, which appeared in the Middle Eocene in America, occur in Morocco by the end of the Eocene, and by the beginning of the Oligocene, in eastern Mesogaea, Madagascar and the Indian Archipelago (Fig. 16.15). Somewhat later, a similar route was followed by the miogypsinids.

Longitudinal separation took place in the Middle Miocene (around 16 million years ago) with:

- the coming together of Africa and Europe in which the temporary emergence of the Straits of Gibraltar interrupted any exchange of fauna; and
- the collision of the Arabian, Indian and Asiatic plates which partitioned Mesogaea into two independent bioprovinces, one Mediterranean and the other Indo-Pacific.

Bioprovinces in the Palaeozoic

Whereas the outlines proposed for the Mesozoic and the Cenozoic are firmly founded, the evidence for the palaeogeography of the Palaeozoic is slender indeed. The schemas put forward rely on the orientation of the pre-Hercynian orogenies and some palaeomagnetic indications, this information being supplemented by the biogeographic application of fossils: archaeocyathids and trilobites, as well as ostracods, conodonts and palynomorphs. The bioprovinces follow a more or less pronounced latitudinal gradient. It is, therefore, possible to given the approximate position of the continental masses (Figs 16.16 and 17) by reconstructing the arrangement of the different sections of the bioprovinces.

Migrations, Barriers and Oceanic Currents

The area of distribution for any species or genus is the result of a complex history, depending both on the mobility of the plates of the lithosphere (longitudinal endemism) and on climatic geographical zonation (latitudinal endemism), as well as on the existing biological possibilities.

Thermal barriers are insuperable. While the sea is no obstacle for planktonic forms, benthic forms of the neritic seas cannot spread without continuity of their biotope. Nevertheless, certain of these – in particular the orbitoidids, nummulitids and miogypsinids – succeeded in crossing the Atlantic in one direction or the other, probably because during their early stage (nucleoconch) their mode of life was planktonic. The manner of their dispersion took different forms: for the orbitoidids of the Cretaceous and the nummulitids of the Eocene, there was a rapid movement from Europe towards America; for the lepidocyclinids and the miogypsinids, there was a slow movement from America towards Europe. This slow migration suggests a series of leaps from one intermediate biotope to another, alternating with long periods of

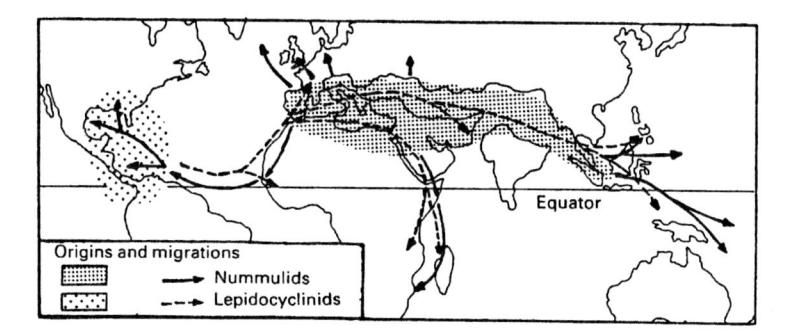

Fig. 16.15 Biogeographies and migrations of some larger benthic foraminifera during the Palaeogene.

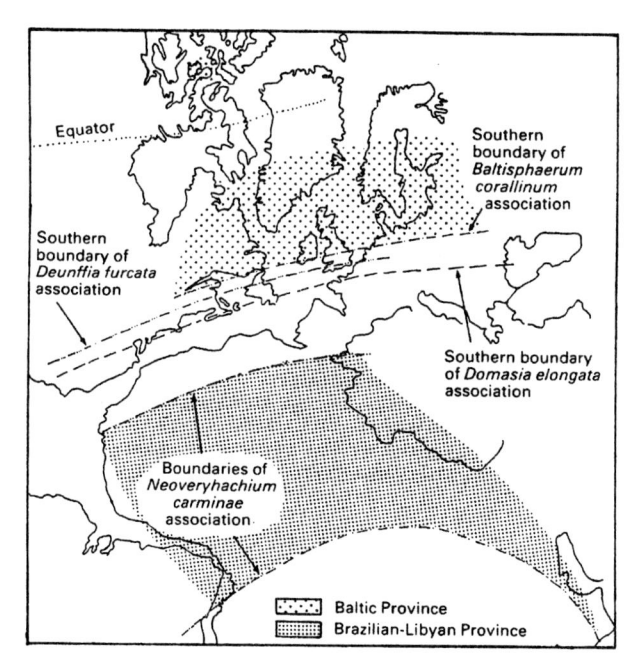

Fig. 16.16 Latitudinal arrangement of areas of distribution of acritarchs and chitinozoans during the Silurian (between 440 and 410 million years ago). After Cramer (1971, fig. 1)

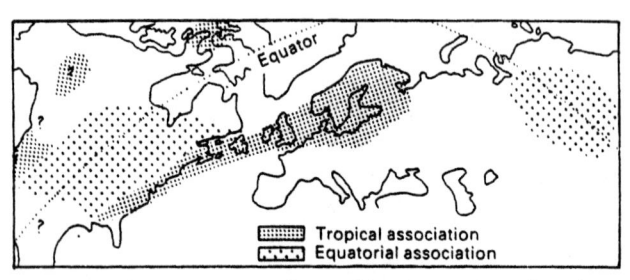

Fig. 16.17 Latitudinal arrangement of areas of distribution of conodonts during the Ordovician (between 510 and 440 million years ago). After Bergström (1973, fig. 5)

colonization during which species evolved where they were. The progress of the migrations seems to be linked with the direction of currents. A proto-Gulf Stream is known to have existed in the North Atlantic at the end of the Cretaceous with a SW–NE orientation and probably compensated by an E-W equatorial current. Movement from Europe towards America would, therefore, have been facilitated. The lepidocyclinids and miogypsinids could not, however, have migrated via the proto-Gulf Stream which ends too far to the north in low temperatures where they could not have proliferated. Transit must have taken place in tropical waters flowing from east to west. For this reason, the dispersion of the organisms was slow, uncertain and discontinuous, and dependent on the occurence in mid-ocean of favourable neritic biotopes: islands and seamounts.

The distribution of benthic foraminifera makes it possible to sketch the palaeocirculation of the surface waters of the oceans. Deep-water currents are approached via abyssal microfossils. Thus, for example, a deepening of the Straits of Gibraltar with a deep current flowing from west to east is attested by a temporary (from 5 to 2 million years ago) influx of psychrospheric ostracods into the western Mediterranean.

Palaeogeography and palaeobiogeography are mutually reliant on each other. In the past, the former greatly assisted the understanding of the latter. In recent times, their roles have been reversed. Established independently of geological and geophysical data, the palaeobiogeography of microfossils assists not only in the reconstruction but also in the dating of stages in the movement of continental masses, the topographical evolution of the sea-floor, climatic changes and ocean current regimes.

CONCLUSION

Being based on the ecology of present-day organisms, the data provided by palaeoecology and palaeobiogeography, although still not conclusive, are leading rapidly to consideration of complex and wide-ranging problems. Factors that make the evidence of microfossils so valuable include:

- Many species are extremely sensitive to ecological variation.
- Being well preserved, their determination is easy.

- They occur in most sediments even if, as with drilling samples, the volume available is small.
- Because of their abundance, statistical methods can be used.

For all these reasons, microfossils are widely employed in attempts to reconstruct the geographies and landscapes that have succeeded each other on the Earth's surface.

BIBLIOGRAPHY

The reconstruction of palaeoenvironments from microfossils is treated in many publications, for example, Reconstruction of marine palaeoenvironments, *Utrecht Micropalaeontology Bulletin* 30 (1983). One example might be the marine palaeoenvironments and continental landscapes of the Paris Basin in the Palaeogene. Works include J.W. Murray & C.A. Wright, *Special Papers of the Paleontological Association of London*, no. 14 (1974); C. Gruas-Cavagnetto, *Mémoires de la Société géologiques de France*, n.s. 56 (1978); and J.-J. Chateauneuf, *Mémoires du Bureau de Recherches géologique et minières*, 116 (1980).

The palaeobiogeography of microfossils is sketched in several articles including: W.A. Gordon, *Bulletin of the Geological Society of America* 81, 1689–1704 (1970); F.C. Dilley, *Geological Journal, Special Issues* 4, 169–190 (1971); C.G. Adams, *Systematics Association Publication*, 7 195–217 (1967). There is a large section on foraminifera in a comprehensive work: A. Hallam, *Atlas of Palaeobiogeography* (Elsevier, Amsterdam, 1973). Also worth consulting is 'Palynologie et dérive des continents', *Sciences Geolgiques, Bulletin*, 27 (1–2) (1974).

Micropalaeontological data will be set in their planetary context by a reading of *Initial Reports of the Deep Sea Drilling Project*, and the work of J.E. Warme, R.G. Douglas & E.L. Winterer, (eds), 'The Deep Sea Drilling Project; a decade of progress', *Society of Economic Paleontology and Mineralogy, Special Publications*, 32 (1981).

Chapter 17
General Conclusion

Apart from the use of the microscope and the need for special preparations, the main distinguishing feature of micropalaeontology is the abundance of the microfossils that form the object of study. Because of this abundance, it is unnecessary to carry out long searches of outcrops for material as a small volume of rock is generally sufficient. These are considerable advantages for field geologists and for those engaged in drilling.

For a long period, the greater part of the activity of micropalaeontologists was devoted, willingly and under pressure of economic requirements (oil exploration), to biostratigraphy. The 'hunt for markers' and the conclusive dating of sedimentary series and geological phenomena has been fruitfully emulated. The results are to be seen in series of microfauna and microflora which are the most accurate chronometers available for Phanerozoic times.

Micropalaeontologists, however, rapidly realized that their goal was not simply a matter of dating. Microfossils are the remains of living organisms and the information that they carry can be correctly interpreted only in the light of their palaeobiology and taphonomy. The need to reconstruct the mode of existence of the microfossils led specialists to collaborate effectively with biologists, oceanographers and sedimentologists in providing a better description of water masses and sedimentary environments.

More recently, the role of micropalaeontologists has been crucial in dating and marking out sea-floor spreading and the movements of the continents. Conversely, oceanographic studies, by gathering materials and adopting new concepts have served to transform micropalaeontology.

The contribution of microfossils is of considerable, and often decisive, importance in earth science, and especially in sedimentary geology and tectonics. It is possible that by participating in the formation of global syntheses and in their confirmation, micropalaeontology may lose its autonomy. This risk, however, is preferable to the baneful and sterilizing effects of isolation and specialization that are a danger in any scientific discipline.

Through their ceaseless investigation of microfossils, micropalaeontologists make a significant and unique contribution to our understanding of the history of the Earth and Life, as well as to the search for materials that are indispensable to mankind.

ACKNOWLEDGEMENTS

Chapter 2
Bignot, G. and Lezaud, L., 1964. *Rev. Micropaléont.*, 7 (2) p. 140.

Chapter 3
Bellemo, 1974. *Bull. Geol. Inst. Uppsala* 6, p. 4.
Bolstovsky, E. and Wright, R., 1976. *Recent Foram.* (Junk, Amsterdam), p. 136.
Hottinger, L., 1977, *Mem. Mus. Nat. Hist. Nat., n.s. (C), Sci. Terre*, 40, p. 39.
Le Calvez, J., 1938, *Arch. Zool. Exp. Gener.*, 80(3) p. 266.
Le Calvez, J., 1950. *Arch. Zool. Exp. Gener.*, 87(4) p. 227.
Loeblich, A.R. and Tapmann, H., 1964. *Treat. Inv. Pal. Protista* 2 (Univ. Kansas Press) p. C99.
Reiss, Z., 1957. *Cont. Cushman Found. Foram. Res.* 8(4) p. 128.
Rhumbler, L., 1959. *In*: Y.A. Orlov (Ed.), *Fund. Pal., Protozoa* (Moscow) p. 132.
Sliter, W.V., 1974. *Lethaia*, 7(1) p. 6.

Chapter 4
Herrig, E., 1966. *Paläont. Abh.*, A, 2 (4) p. 739.
Kesling, R.V., 1951. *Illinois Biol. Monogr.*, 21(1/3), pp. 6–7.
Martinson, A., 1962. *Bull. Geol. Inst. Univ.* (Uppsala), 41, pp. 80 and 83.
Morkhoven, F.P., 1962. *Post-Palaeoz. Ostr.* (Elsevier, Amsterdam) p. 6.

Chapter 7
Gaarder, K.R. and Markali, J., 1956. *Nytt. Mag. Bot.* (Oslo) p. 1.
Lecal-Schlauder, J., 1951. *Ann. Inst. Oceanogr.* 26, p. 307.

Chapter 8
Bergon, P., 1974. *In*: F. Oltmanns (Ed.) *Morph. Biol. Algen* (Asher, Amsterdam) 1, pp. 133, 189 and 193.
Gran, H., 1933. *In*: P. Dangeard (Ed.) *Traité d'Algologie* (Lechevallier, Paris) pp. 106.
Pfitzer, E., 1974. *In*: F. Oltmanns (Ed.) *Morph. Biol. Algen* (Asher, Amsterdam), 1, p. 133.

Chapter 9
Lindström M., 1973. *Geol. Soc. Am., Spec. Pap.*, 114, pp. 18, 20 and 88.
Rhodes, F.H., 1954. *Cambridge Philos. Soc. Biol. Rev.*, 29, pp. 419–452.
Seddon, G. and Sweet, W.C., 1971. *J. Paleontol.*, 45(5) p. 873.

Chapter 10
Erdtman, G., 1963. *Handbook of Palynologym* Copenhagen (Munksgaard, Copenhagen) pp. 23–5.
Evitt, W.R., 1967. *Stanford Univ. Publ. Geol. Sci.*, 10(3), p. 15.
Evitt, W.R., 1969. *In*: R.M. Tschudy and R.A. Scott (Eds.) *Aspects of Palynology* (Wiley, New York) pp. 144, 446 and 458.
Grebe, H., 1971. *C.I.M.P., Microf. Org. Paleoz.* (C.N.R.S., Paris) 4, Spores, p. 27.
Koslowski, R., 1963. *Acta Palaeont. Polonica*, 8(4) pp. 433 and 435.
Paris, F., 1981. *Mem. Soc. Geol. Min. Bretagne*, 26, pp. 68 and 75.
Sarjeant, W.A., 1974. *Fossil and Living Dinoflagellates* Academic Press, London, p. 66.
Van Campo, M. and Sivak, J., 1972. *Pollen and Spores*, 14(2) pp. 135 and 137.
Wall, D. and Dale, B., 1968. *Micropal.*, 14(3), p. 267.

Chapter 13
Barghoorn, E.S. and Tyler, S.A., 1965. *Science*, 147, pp. 567-571.
Bé, A.W. and Tolderlund, D.S., 1971. *In*: B.M. Funnel and W.R. Riedel (Eds.), *Micropal. Oceans* (Cambridge Univ. Press, Cambridge) pp. 109 and 117.
Benson, R.H., 1973. *Lethaia*, 8(1) p. 79.
Blow, W.H., 1956. *Micropal.*, 22(2) p. 68.
Buchner, P., 1963. *In*: V. Pokorny (Ed.), *Princ. Zool. Micropal.*, p. 307.
Caron, M., 1967. *Eclog. Géol. Helv.*, 60(1) p. 60.
Casey, R.E., 1971. B.M. Funnel and W.R. Reidel (Eds.), *Micropal. Oceans*, p. 157.
Gerhardt, H., 1963. *Boll. Soc. Palaeont. Ital.*, 2(2) pp. 61.
Hottinger, L., 1960. *Mém. suisses Paléont.*, 75/76, pp. 50, 136 and 143.
Jannin, F., 1967. *Rev. Micropal.*, 10(3) p. 167.
Kinne, O., 1970. *Marine Ecology*, (Wiley, London) Vol. 2 p. 824.
Magniez-Jannin, F., 1975. *Cah. Paléont.*, pp. 118–119.
Myers, E.H., 1943. *Amer. Philos. Soc. Proc.*, 86(3) p. 444.
Peypouquet, J-P., 1975. *Bull. Inst. Géol. Bassin Aquitaine*, 17.
Pflug, H.D., 1976. *Naturwiss.*, 54, p. 238.
Pflug, H.D. and Jaeschke-Boyer, H., 1979. *Nature*, 280, p. 484.
Rauzer-Chernousova, D.M., 1963. *Evol. trends Foraminifera* (Elsevier, Amsterdam), p. 48.
Schopf, J.W., 1968. *J. Paleont.*, 42(3) pp. 680 and 682.
Schroeder, R., 1975. *Rev. Esp. Micropal.*, (Sp. Issue) p. 123.

Chapter 14
Arrhenius, G., 1963. *In*: M.N. Hill (Ed.), *The Sea* (Interscience, New York) Vol. 3, p. 657.
Bourrelly, P., 1970. *Algues d'eau douce* (Boubee, Paris), 3, p. 447.
Duparque, A., 1933. *Mém. Soc. Géol. Nord*, 11
Gebelein, C.D., 1969. *J. Sed. Petr.*, p. 60.
Walter, M.R., 1972. *Palaeont. Assoc. London, Spec. Publ.*, 11, p. 144.

Chapter 15
Bignot, G. and Cavelier, C., 1981. *Bull. Inf. Geol., Bassin Paris*, Mem. h.-s. 2, p. 8
Bolli, H., 1957. *US Nat. Mus. Bull.* 215, *St. Foramin.*, p. 63.
Bolli, H., 1966. *Bol. Inf. Asoc. Venezolana Geol., Min. Petr.*, 9(1) p. 32.
Cavelier, C., 1979. *Sc. Geol.*, 54, p. 68.
van Vlerk, I.M., 1968. *Micropal.*, 14(3) pp. 335 and 336.

Chapter 16
Arni, P., 1965. *Mém. B.R.G.M.*, 32, p. 18.
Barr, F.T., 1962. *Paleont.*, 4(4), p. 560.
Bergström, S.M., 1973. A. Hallam (Ed.), *Atlas of Palaeogeography* (Elsevier, Amsterdam) p. 55.
Blance-Vernet, L., 1972. D.J. Stanley (Ed.), *The Mediterranean Sea* (Dowden, Stroudsberg) p. 119.
Caralp, M. and Pujol, C., 1974. *Bull. Inst. Géol. Bassin Aquitaine*, 16, pp. 35 and 46.
Cramer, F.H., 1971. *Mém. B.R.G.M.*, 73, p. 231.
Elhai, H., 1963. *La Normandie occidentale. Etude morphologique* (Bière, Bordeaux) p. 624.
Grimsdale, T.F. and van Morkhoven, F.P., 1955. *Proc. 4th World Petrol. Congr.*, 4, p. 475.
Haq, B.U., Okada, K. and Lohmann, G.P., 1979. *In. Rep. D.S.D.P.*, 43, p. 622.
Heirtzler, J.H., Dickson, G.O., Herron, E.M., Pitman, W.C. and Le Pichon, X., 1968. *J. Geophys. Res.*, 73, pp. 2119–2136.
Muller, C. Sigal, J., 1979. *Bull. Song R. Schlich, Geol. France*, 21(1) pp. 57–63.
Murray, J.W., 1973. *Distribution and Ecology of Living Benthic Foraminifera* (Heinemann, London), p. 241.
Shackleton, N.J. and Kennett, J.P., 1975. *In. Rep. D.S.D.P.*, 29, p. 47 and 748.
Worsley, T. and Martini, E., 1970. *Nature*, 225, p. 1243.

Index

abdomen (radiolarian) 88
acme 178
acritarchs 6, 17, **121-123**, 127, 128, 142, 203
acrozones **178**, 179
agglutinated structure **22-23**, 27, 29, 32, 37, 41, 58, 131, 140, 141, 158, 191
air sacs 110
Albaillellaria 84, 88, 97
Alcyonaria 67, 68, 131, 134
algae 2, 44, 59, **61-65**, 71, 72, 89, 94, 96, 115, 125-126, 138, 171, 172, 174, 188
alveoli
 diatoms 91
 foraminifera 24
 pollens 109-110, 115
Alveolinidae **34-35**, 37, 39, 41, 153, **160-162**, 189, 201, 202
Ammodiscidae 32, 41
ammonites 88, 153, 183, 188
'ammonitico rosso' 171
ampulla 123
anagenesis 156, 157, 159
angiosperms 109, 110, 114, 130, 137
annelids 124, 126, 127, 138
annular tests 29
anoxic environments 123, 133, 163, 167, 173, 174
anteturmas 112
antheridum 107
antherozoids 107
aperture
 calpionellids 56
 foraminifera 27
appendages
 dinocysts 117
 ostracods 43, 44
aragonite 23, 34, 61, **65-71**, 134-137, 171, 172-173
Aragonite Compensation Depth (ACD) 136
araphids 91, 93
Archaean 165, 166, 167

archaeocopids 51, 54
archaeomonads 96
archaeopyle 80, **118**, 119, 120
Archaeoraphids 91, 93
arenaceous tests 23
Arkhangelskiellaceae 76
arthropods 43, 131
Ascidia 68, 70, 131, 133, 134, 168
assemblages 102-103
Atlantic 200-202
auxospores 89, 90
axopodia 84

Bacillariophyceae see diatoms
bacteria; bacteriomorphs 129, 130, 133, 138, 163, 165, 166, 167, 171
Bairdaceae 51, 53, 54
basal cavity 99-100
basal plate 99-100
basal ring 94
Bennettitales 115, 126
benthic; benthos 22, 45,54, 59, 66, 74, 88, 90, 93, 140, **146-147**, 150, 190-191, 192, 196, 200-201
beyrichicopids 49
binocular lens see microscopes
binomen 16
bioclastic sedimentation 169-172
biocoenoses 149-151
bioprovinces 150, 197, **199-202**
biostratigraphy, 180, **181-185**, 206
biostystematic descent 151
biotopes **145-146**, 149-151, 187, 195, 202-204
bioturbation 140-141
biozonation 177, **178-181**
birds 174
'blooms' 174
Botrytococcaceae 125, 127, 130, 147, 171
Brachiopods 59, 67, 68, 131, 134
brachytelia 160, 162
Branchiopods 43, 51, 131

211

brood pouch 47
Bryophytes 107, 114, 130
Bryozoa 2, 61, 69, 104, 131, 134
Buliminidae 34, 41

calciodinellids 80-81
Calciosoleniaceae 76
calcispheres 59, 129, 130
calcisponges 68, 131, 134
Calcite Compensation Depth (CCD) 135-136, 169
calpionellids 6, 56-58, 80, 88, 129, 131, 153, 162, 200
calpionellomorphs 58, 59, 131
Cambrian 37, 42, 51, 54, 63, 68, 81, 97, 104, 114, 122, 127
canals 102
caprinid rudists 68
capsular membrane 84
carapace 43, 44, 45-49
Carboniferous 37, 41, 43, 53, 54, 59, 62, 63, 65, 68, 81, 97, 104, 112, 114, 122, 127, 159, 173
cavate cyst 118
cenozones 178
Centrales 89, 91, 94, 97
cephalis 86-87
Cephalopods 59, 68, 131
Chaetognatha 104
chalks 80, 142, 171, 172, 183, 190
chamaeraphids 91, 93
chambers
    Chitinozoa 123
    foraminifera 20, 22, 26-30
    radiolarians 88
channels (of hyaline tests) 25
chitinoid basal layer 23, 24, 25, 126, 131
chitinoid scolecodonts 126, 131
Chitinoidella 57
chitinous remains 46, 56, 131
Chitinozoans 6, 123-125, 126, 127, 129, 130, 143, 180, 183, 203
Chitins (in palynomorphs) 106, 133
Chlorococcales 125, 127, 130
chloroplasts 115
Chlorophytes 61-62, 125, 130, 190
chomata 29
chorate cyst 117
chordates 104
chronostratigraphy 180, 182-185
chronozones 180
Chrysomonadales 72, 94-96, 97
Chrysophytes 72, 89, 130
cingulum 115
cirripeds 43, 68, 131, 134
cladogenesis 156, 158, 160
clays 6, 65
coal 6, 106, 125, 172, 173, 185
Coccolithidae 76
coccolithophores 72-74, 129, 130, 136, 137, 145, 147
coccolith ooze 169-170
coccolith zones 180, 181, 183, 186, 187, 195, 200-202
coccoliths 1, 5, 24, 57, 72-80, 88, 96, 130, 134, 135, 139, 142, 153
coccospheres 72, 73, 74, 75, 76, 130
collar
    calpionellids 56
    Chitinozoa 123
collection 5-7
colorimetric stages 143
colporous pollens 109
colpus 109
conchostraceous brachiopods 43
concurrent range zones 178
confinement 188
Coniferales 115, 130, 140
Conodont Alteration Index (CAI) 143
conodont zones 180, 183, 185
conodontophorids 103-104, 168
conodonts 1, 6, 8, 88, 99-105, 129, 131, 133, 142, 180
continental drift 197-198
copepods 43, 84, 131, 135
coproliths 131, 138
copula 123
coral reefs 171, 174
corallinaceous rhodophytes 62, 63, 69, 130, 134, 171, 189
corals 68, 88, 169, 171
Cordaitales 115
Cretaceous 171, 172, 197-198
    calcareous microfossils 56-58, 59
    calcareous nanofossils 76, 80, 136, 183, 201
    foraminifera 37, 41, 136, 154, 155, 160, 180, 182, 183, 190, 192, 201
    ostracods 53-54
    palynomorphs 111-126
    siliceous microfossils 88, 93-96
Cricoconarids *see* Tentaculids
cricoliths 75
crinoids 61, 67, 68, 102, 131
cross-lamellar molluscs 69
crustaceans 43, 131, 138
cryptomerial Rhodophyta 63, 131
cuticular debris 130, 173
cutins 106
Cyanophytes 61-63, 130, 134, 152, 163, 166, 167, 174-176
Cycadeles 115, 130
cyclic tests 29
cycloliths 75
Cypridaceae 45, 51, 53, 54
cysts 122, 124
    diatoms 89, 90
    foraminifera 20, 22
    siliceous 95, 96, 130
    see also dinocysts

INDEX 213

Cytheraceae 46, 53, 54

Darwinulaceae 49, 51, 53
Dasycladales 59, 61, 69, 70, 130, 134, 189
Deep Sea Drilling Project (DSDP) 198, 199, 205
description 15-17
Devonian 37, 41, 53, 54, 57-59, 63, 65, 66, 69, 70, 97, 103, 104, 111, 112, 113, 124, 126, 127
dextral spirals 29, 64, 65, 149, 192
diagenesis 59, 69, 75, 106, 136, 142, 163, 172, 197
diaphanothecate tests 25, 34
diaphragms 91, 93
diatomites 96, 146, 171,
diatoms 6, 22, 84, 89-94, 97, 130, 134-137, 147, 169, 174, 197
dicaryon 115
dicotyledons 108, 115, 130
Didemnidae 66
dimorphism 21, 37, 44, 47, 49, 84-86
dinocysts 6, 59, 80-81, 117-121, 126-127, 130, 143, 145, 184, 185, 191
dinoflagellates 1, 79, 80, 96, 107, 115-116, 127, 130, 173, 174
Dinophysiales 119, 120, 130
diploid sporophyte 107
Discoasteridae 78, 80, 81
discoliths 74, 75, 76
Discorbidae 34, 41
displacement, natural *see* bioturbation
Dogger 68, 119
dolomites 6, 66, 83, 102, 163, 189

ebridians 95, 96, 97
echinoderms 59, 61, 67, 68, 131, 134
echinoids 68, 131, 183
ecomorphs 146-147, 192
ecostratigraphy *see* biostratigraphy
ecosystems 149; *see also* environments
ecotypes *see* ecomorphs
ecozones 189
ectexine 110
ectoderm 123
ectoplasm
 foraminifera 19
 radiolarian 84
electronic probe microanalyser 15
embryont *see* juvenorium
endemism *see* bioprovinces
endexine 110
endophragm 118
endoplasm
 foraminifera 19
 radiolarian 84
endosperm 123
endospores
 diatoms 89, 94
 spores 110
Endothyridae 32, 37, 41

Entactinids 84, 88
environments; palaeoenvironments 146-151, 187-205
Eocene 53, 63, 66, 71, 94, 121, 142, 160, 161, 174, 177, 181, 184, 189, 196, 202
epibole 178
epigenesis; epigeny 83, 137
epitheca 89
Equisetaceae 96
ethology 146, 153
Eufilicales 107
euraphids 89, 91, 93
Euripterids 126, 131
eurybiota 146
euryhalines 54, 146, 148
evaporites 6, 189
evolutionary trends 159-162, 201
exine 109-115
exoexine 110
extraterrestrial life 167

fibroradial structure 24, 25, 37, 56, 58, 59
'filaments' 63, 69, 70, 131, 163-165, 174-176
filipodia 84
fish 67, 68, 102, 131, 134, 174, 188
flagellae 115
foramina 20
foraminifera 2-3, 6, 15, 19-42, 126, 131, 134-142, 145-153, 157, 158, 160, 161-162, 174, 177, 179-182, 184, 187-193, 196-197, 200-202, 204
fossil fuels 174; *see also specific types*
francolite 99
frustules 89, 90-94, 171
fungi 130, 138, 173; *see also* hyphae
Fusilinidae 29, 32-34, 41, 102, 153, 159, 160

gametangia 65
gametophyte, haploid 107
gamont 21, 22
gastropods 61, 65-66, 124, 131, 138
genus; subgenus 16
germinal aperture 107
Gigantostraceae 126, 131
Ginkgoales 115, 130
girdle 89
glauconites 8, 102, 123, 137, 183
Globigerinidae 34, 41
gnetales 130
gnetophytes 115
goniatites 88
gonyaulacean lineage 121
Gramineae 96
granular ectexine 110
granular tests 24, 25
graptolites 88, 102, 123, 124, 131, 183, 188
Gymnodiniales 95, 96, 116, 119, 120, 122, 130
gymnosperms 108, 110, 114-115
gyrogonites 56, 64, 65, 130, 188

haploid gametophyte 107
haptonema 72
helioliths 75
heterococcoliths 75
heterospory 107, 112, 114
hexactinellid sponges 68
hinge 43, 45, 47
Holocene 194
holococcoliths 73, 75
holothuroids 67, 68, 70, 131, 133, 168
hyaline tests 23, 25, 27, 29, 34, 37, 41, 134, 141, 191
Hydrodictaceae 125, 127, 130
Hydrofilicales 107, 190
hyperhaline environments 150, 152
hyphae, mycelian 163; see also fungi
hypohaline environments 150, 152
hypothallus 63
hypotheca 89, 93
hystrichospheres 117, 119, 120, 128, 145

imperforate tests see porcelaneous tests
inaperturates 108, 112, 114, 115
intercalary series 116
International Phase of Ocean Drilling (IPOD) 198
interval zone 178, 179
intexine 110
intine 110
isopory 107
isotope ratios 15, 148
isotopic analysis 194-197

Joint Oceanographic Institutions Deep Earth Sampling (JOIDES) 198
Jurassic 37, 41, 53, 54, 58, 63, 68, 69, 71, 79-82, 88, 97, 119, 122, 127, 162, 174, 197, 200
juvenarium 26, 34, 38, 39

keriothecate tests 25, 34
kerogen 133, 173-174

lagoons 22, 189, 191
lamellar structure 25, 37, 46, 47, 58, 66, 69, 70, 99-102
Lamellibranchs 45, 131
lamina 91
Leiosphaeridiacae 125, 126, 127, 130
lenses see microscopes
leperditicopids 45, 53, 54
lepidocyclinids 202, 203, 204
Lias 53, 54, 76, 80, 114, 121, 126, 180
ligament 43
lignins 106, 125, 130
lignite 6, 189
lignocellulosic cements 172
limestones 6, 9, 45, 53, 59, 65, 80, 102, 138, 171, 173, 189
lineage zones 181-182

lithistids 66, 131
lithogenesis 169-176
Lithothamniae 63
Lituolidae 24, 30, 41, 200, 201
lopadoliths 74, 76
loricas 56, 58, 59, 126, 131
Lycopodiales 107, 112, 130
lysocline 135-136

macrobiota 1
macrospores (see megaspores)
madreporaria 69, 70, 131, 134; see also corals
magnetic anomalies 198-200
magnifying glasses 2, 5, 10
'majolica' 171
mammals' teeth 67, 68
marbles 6
'mare spoco' 174
marls 6, 9, 45, 75, 80, 141, 172, 189
megalospheres (form A) 20, 21, 22, 38, 155, 158
megaspores 107, 110, 111, 173; see also spores
Metacopida 53, 54
metamorphic rocks 6, 142, 163-167, 185
meteorites 167
microanalyser, electron probe 15
microbiota 1
Microcodium 130, 141
microfacies 14, 15, 70, 177-178, 185
microforaminifera 126, 127; see also chitinoid basal layers
microgranular structure 25, 32, 37, 41, 56, 58, 59
micropalichnology 138
microprobe, electron 15
microrhabulids 78-80
microscopes 2-3, 10-14, 15, 18
  binocular lens (stereomicroscope) 8, 10-11, 14
  electron 1, 12-13, 75
  metallographic 173
  optical (light) 10-11, 18, 69, 70, 74, 76, 79, 91, 171
  polarising 15, 75
  scanning electron (SEM) 12-13, 14, 16, 69, 78, 94, 171
  transmission electron (TEM) 12, 14, 110
microspheres (form B) 21, 22, 133, 155, 156, 162
microspores 107-108, 173; see also spores
miliolids 22, 29, 34, 37, 42, 141, 162, 189
Miocene 39, 94, 96, 119, 160, 182, 196, 202
miogypsinids 34, 37, 41, 202-204
miospores 107; see also microspores
molluscs 61, 66, 69, 134, 171, 174, 195
monocotyledons 108, 115, 130
monolete spores 107, 108, 110, 114
monoraphids 91, 92
monosulcate pollens 108, 114, 115
morphofunctional analysis 153-159
muds and mudflats 134, 136, 138

INDEX

muscle scar field 47
myodocopids 45, 51, 53, 54

Nannoconus 57, 80, 88, 129, 130, 153, 171, 200
nanoagorites 171
nanofacies 14, 70, 75, 78, 172
nanofossils 1, 6, 7, 11, 14, 72-82, 107
Nasselaria 84, 86, 88, 97, 139
nektonic; nektos 88, 98, 123, 147, 188
nematodes 138
Neogene 41, 54, 66, 81, 88, 97, 114, 127, 192
Nodosariidae 34, 37, 41, 152
nomenclature 16-18, 75
Normapolles 112, 114, 115
nucules *see* gyrogonites
Nummulitidae 2, 3, 25, 34, 37, 39, 134, 151, 189, 202, 203

observation 11-15
oceanicity 188, 191
oil 2, 185, 206
Oligocene 39, 54, 80, 94, 115, 182, 185, 196, 202
oncoliths 130, 174
ontogenesis 22, 47, 48, 162
oogones *see* gyrogonites
oospheres 107
oozes, oceanic 66, 136, 138, 169-171
opal 83, 86, 89, 137
operculum
   Chitinozoa 123
   dinocysts 119
ophiurids 67, 68, 131
Opisthobranchia 65
orbitoid tests 30
Orbitoididae 30, 34, 37, 41, 201, 202
Orbitolinidae 32, 41, 157, 182, 202
Ordovician 37, 41, 53, 54, 59, 81, 88, 97, 104, 114, 122, 124, 126, 127, 180, 204
origin of life 163-168
ornamentation 30
orthogenesis 162
ortholiths 75
ostracods 3, 6, 17, 43-55, 88, 131, 134, 147-148, 152, 156, 177, 188, 197, 202, 204
otoliths 67, 68, 131, 134

Palaeocene 71, 80, 119, 141, 161, 162, 184, 185, 195, 196
palaeocopids 49, 50
palaeoenvironments *see* environments
Palaeogene 41, 54, 75, 81, 94, 114, 127, 177, 203, 204
palaeogeography 200-205
palaeosols 68, 96
palaeotemperatures 191-197
palichnology *see* micropalichnology
palingenesis 162
palynofacies 107

palynology 3, 6, 9, 12, 14, 18, 106-128, 140, 163, 173, 189, 194
palynomorphs 101, 117-121, 126, 129, 133, 142-143, 173, 190, 202
parachomata 29
paraconodonts 104
parataxonomy 75, 102, 112
Parathuramminidae 34, 59
peats 3, 96, 134
pelagic; pelagos 56, 59, 65, 69, 104, 124, 146-147
Pennales 90, 91, 93, 94, 97
perforate tests *see* hyaline tests
perforations 57, 58, 134, 138
pericoel 122
periderm 123
peridinacean lineage 121
Peridiniales 116, 117, 119, 130, 137
periphragm 118
perispore 110
perithallus 63
Permian 41, 54, 68, 81, 97, 104, 114, 119, 126, 127
phosphatic fossil remains 8, 45, 67, 68, 69, 83, 99-105, 131, 137, 138, 143
phosphatic sediments 123, 173
phylozones 181-182
phytoliths 95, 96, 97, 131
pillars (foraminifera) 29
placoliths 74, 76
planispiral tests 27, 28, 29, 32, 34, 38
planktonic; plankton 22, 88, 96-97, 134-136, 141, 144, 146-150, 153, 178-185, 188-197, 199, 202; *see also* planktonic species
planktonic species
   acritarchs 123
   calcareous nanofossils 74, 82
   calpionellids 56-57
   calpionellomorphs 59
   chitinozoans 123
   conodonts 102
   diatoms 89, 93
   dinocysts 121, 174
   dinoflagellates 116, 174
   ebridians 96
   foraminifera 34, 37, 39, 41, 139, 153, 160, 162, 169, 179, 180, 181, 187-197, 200-201
   isolated organic elements 67
   ostracods 45, 49, 54
   radiolarians 86, 139
   silicoflagellates 94
   tentaculids 66
plants, higher 96, 107
plates *see* cirripeds; conodonts; dinocysts; frustules; thin sections
platform conodonts 99, 100, 104
Platycopina 49, 51, 53, 54
Pleistocene 97, 177, 194, 196
Pliocene 78, 80, 196

podocopids 44, 46, 49, 51, 53, 54
pollens 6, 72, 101, 106-115, 127, 128, 137, 140, 143, 167, 177, 188, 191, 194
polychaeta 126, 131, 138
Polycitoridae 66
polycystine radiolarians 84-88, 93, 134, 151
polyplicate pollens 112, 114, 115
porcelaneous tests 23, 24-25, 27, 29, 34, 41, 141, 191
pores
    diatoms 89, 91
    ostracods 47
    radiolarians 88
postabdomen 88
Prasinophycease 125, 130
Precambrian 104, 126, 127, 163, 166, 174
preparation 7-10
prepollens 112, 114, 130
'primeval soup' 163, 166, 167
prismatic molluscs 69
productivity, organic 197
productivity, planktonic 136
proloculus 26
prosome 123
Proterozoic 163-165, 166
protobiota 163, 166, 167
provinces 150; see also bioprovinces
proximate cyst 117
pseudochitinous capsular membrane 84
pseudofibrous tests 25, 32, 37
pseudorbitoids 201
pseudopodia
    foraminifera 19
    radiolarians 84
pseudoraphe 91, 93
pseudostome 123
Psilophytineae 112, 130
psychrosphere 150, 152, 204
Pteridophyta 107, 112, 114, 115, 130
Pteridosperms 112, 130
Pterobranchia 104, 131
pteropods 61, 65-66, 129, 131, 134-137, 147, 170
punctae 91
Purbeckian facies 3, 53, 188
pylome 121
pyrites 9, 45, 83, 137

quartz; quartzites 142, 166, 167; see also siliceous rocks
Quarternary 3, 66, 68, 123

radiolarians 6, 59, 80, 84-88, 94, 97, 134-137, 139, 147, 151, 169, 170, 172, 185, 188, 197, 200
radiolarites 6, 172
radiolitid rudists 69
raphe 89, 91, 93
'red tides' 116, 174

reefs coral see coral reefs
reworking 139, 142
rhabdoliths 74, 76
Rhabdosphaeraceae 76
Rhodophyta 63, 71, 130, 174
rhyncolites 67, 68, 131
Rotaliidae 34, 37
Rotaliina 34
rubbing 7, 14, 74, 78
Rudistaceae 69, 88

saccates 110, 112-115
salinity 146-148, 152, 188-191
sands and sandstones 5, 9, 66, 102, 123, 171
saprocols 171
sapropels 136, 171, 173
scapholiths 76
schales 6, 9, 102, 125, 142, 163, 165, 166, 174
schizont 21, 22
Schizosphaerella 78, 79, 81
sclerites 67, 68, 71, 131, 133, 134
sclerodermites 69
scolecodonts 126, 127, 131
sculptural elements 110
Scyphosphaeraceae 76
sea-floor spreading 198-200
Selaginellales 107
SEM see microscopes
septa
    foraminifera 26
    radiolarians 88
septulae 29
serpulids 59, 69, 131, 134
setae 47
silica; silicates; silicon 1, 9, 45, 148, 173
siliceous microfossils 68, 72, 83-98, 130, 131, 134, 136, 146, 163, 166, 197
siliceous rocks 6, 8, 65, 102, 142, 165-171, 172
silicisponges 58, 94, 97, 131, 134
silicoflagellates 94-96, 97, 98, 130, 134-137, 145, 147, 185, 191
Silurian 37, 41, 48, 54, 59, 66, 81, 97, 104, 112, 114, 119, 122, 124, 127, 180, 203
sinistral spirals 29, 64, 65, 149, 192
skeletal fragments 68-71
skeletons
    radiolarians 86-88
    silicoflagellates 94-96
'slides' 8; see also washing
Solenoporaceae 63, 130
Soritidae 34
sorting 8, 11
speciation 145, 153-159, 168, 200
species; subspecies 16
Spermatophyta 107, 110
spicules 24, 66, 67, 68, 84, 86, 94, 96, 131, 133, 172
spines
    dinocysts 117

INDEX

radiolarians 86
spirals *see* dextral; planispiral; sinistral; streptospiral; trochospiral sponges 24, 61, 66, 131, 133, 172; *see also* calcisponges; silicisponges
spongoliths 172
sporangium 107
spores 6, **106-115**, 127, 128, 130, 137, 140, 143, 147, 173, 188
sporoderm 109
sporomorph 106
sporopollenins 80, 106, 110, 117, 125, 133, 143
Spumellaria 83, 84, **86**, 88, 97, 139
statospores *see* endospores
stenobiota 146
stenohalines 22, 23, 54, 86, 147, 188
stereomicroscope *see* microscopes
stolons 27
stratotypes 183-185
streptospiral tests 29-30
striated pollens 112, 114
stromatolites 63, 130, 163, 165, 166, **174-176**, 189
stromatoporoids 69, 131
submergence, tropical 150, 151
subturmas 112
sulcus
  dinoflagellates 115
  pollens 109
synonymy 16
systematics 153
Syracosphaeraceae 76

tabulation 81, **116-119**, 122
tachygenesis 22
tachytelia 160, 162
taphocoenoses 68, 138, 144, 151, 153, 187, 188, 191, 192
tasmanine 125
Tasmanitaceae 125, 127, 130
taxa; taxonomy 75, 153, 159, 162, 178, 179, 180, 181
tectate pollens 110
tectoria 29
tegillum 27
TEM *see* microscope
tentaculids 2, 61, 65, 66, 71, 129, 131
tests *see* foraminifera; radiolarians; silicoflagellates
Tethys 200-202
tetrads 107
Textulariina 32
thallus **61-63**, 130, 141, 171
thanatocoenoses 140, 141, 142, 190
theca
  diatoms 89
  dinophyceae 116, 117
Thecosomata 65
thermosphere 150

thin sections 3, 6, **9-10**, 14, 30, 43, 45, 56, 67, 68, 70, 173
thorax 88
tintinnids 57, 126, 127, 131, 147
tooth (of test) 27; *see also* mammals' teeth
trabeculae 86
tremaliths 74
trematophore 27
Triassic 37, 41, 43, 53, 54, 76, 78, 88, 97, 99, 104, 111, 112, 114, 119, 127, 138, 180
trichomes 63
trilete spores 107, 108, 110, 112, 114, 115
trilobites 61, 88, 131, 183, 202
trochospiral tests 27, 28, 29, 32, 34, 36
tubercule 43
turmas 112

Utodeaceae **61-63**, 69, 130, 134, 171
umbellinids 64, 65, 129, 130
umbilicus 29
upswellings 136, 150
utriculus 65

valves
  diatoms 89, 91, 93
  foraminifera 27
  ostracods 43, 45
variability, morphological 17, 154-155
vautrinellids 59
vertebrates 99, 103, 104, 166
vestibule 46-47
vesicles
  Chitinozoa 123
  conodonts 102
vitrinite 173

washing and wash residues 7-8, 14, 43, 55, 67, 68, 70, 126
Wealdian facies 53, 188, 198
'white matter' 100, 102
windows 91

zones *see* biozones; chromozones; ecozones; lineage zones; phylozones
zooanthellae 116
zygoliths 75

Printed in the United Kingdom
by Lightning Source UK Ltd.
102188UKS00003B/136-138